MATH
LESSONS FOR A LIVING EDUCATION

Teaching Companion

MASTER BOOKS
— CURRICULUM —

Master Books Creative Team:

Editors: Craig Froman

Laura Welch

Design: Terry White

Cover Design: Diana Bogardus

Copy Editors:

Judy Lewis

Willow Meek

Curriculum Review:

Kristen Pratt

Laura Welch

Diana Bogardus

Specials thanks to the families and students who allowed their photos to be used in this *Teaching Companion*. All other images are istock.com.

First printing: January 2020
Second printing: August 2020

For information write:

Master Books®, P.O. Box 726, Green Forest, AR 72638

Master Books® is a division of the New Leaf Publishing Group, Inc.

ISBN: 978-1-68344-215-8
ISBN: 978-1-61458-733-0 (digital)

Scripture quotations marked NLT are taken from the *Holy Bible, New Living Translation*, copyright © 1996, 2004, 2015 by Tyndale House Foundation. Used by permission of Tyndale House Publishers, Inc., Carol Stream, Illinois 60188. All rights reserved.

Scripture quotations marked NKJV are taken from the New King James Version®. Copyright © 1982 by Thomas Nelson. Used by permission. All rights reserved.

Printed in the United States of America

Please visit our website for other great titles:
www.masterbooks.com

"Have not I commanded thee?
Be strong and of a good courage;
be not afraid, neither be thou dismayed:
for the LORD thy God is with thee
whithersoever thou goest" (Joshua 1:9; KJV).

As a homeschooling mom and author, **Angela O'Dell** embraces many aspects of the Charlotte Mason method yet knows that modern children need an education that fits the needs of this generation. Based upon her foundational belief in a living God for a living education, she has worked to bring a curriculum that will reach deep into the heart of home-educated children and their families. She has written over 20 books, including her history and math series. Angela's goal is to bring materials that teach and train hearts and minds to find the answers for our generation in the never-changing truth of God and His Word.

Table of Contents

Dedication: To all of the parents who are answering the call of God to educate their children at home. I am so very proud of you. *Soli Deo Gloria.*

Also dedicated to the Author and Sustainer of all things, Jesus Christ, the Word Who became flesh to dwell among men, through whom all of us have the privileged ability to approach the very throne room of God.

"If Jesus were to release His sustaining power, even for a moment, our universe would fly apart in chaos. The only reason math works is because Jesus created it from His mind of order and precision. Randomness and chance never produce predictable constancy and accuracy."
                             —Israel Wayne, *Education: Does God Have an Opinion?*

Start Here

Important Note to the Homeschool Parent

Dear Homeschooling Mama, Dad, or Grandparent, welcome!

When I decided to make this resource for you, I wanted to make sure I was giving you useful and practical teaching tips that would help make your day run more smoothly. I know from experience how foggy a mom's brain can get when she's juggling multiple children at various levels of development. I also wanted this little book to be comforting to your soul, a voice of truth and peace to help calm any jitters you may have about your own personal journey of learning to facilitate your child's education. I don't want you to try to keep the pages all clean and neat; this book is meant to be used as a resource…one which will eventually sport worn and highlighted pages, marked by a hundred colored tabs and dog-ears.

So here is what I want you to do: wait until your children are napping — or better yet, down for the night, pour yourself a cup (mine's coffee, although tea may be better in the evening), put your feet up, and take some time to walk with me through this book. **Please keep in mind that it would be highly beneficial — and make more sense — to read this book in its entirety instead of jumping ahead to a specific section.**

Before we begin our journey through the rest of this *Companion*, there is one general topic I want to talk to you about right at the beginning — something about which I hold a strong conviction.

The secular worldview has drilled into our heads and hearts that we as parents do not have what it takes to teach our children. I will go into depth about how we can stand against this mindset in a later section of this *Companion*, but for now, I want to outline my personal convictions concerning home education and my role in your life. I know from experience that as human homeschooling parents, it is easy for us to always be looking for a foolproof approach — a guarantee that our efforts are going to pay off and that we will not fail in our colossal efforts. We all have the constant reminder in the back of our mind that "the rest of the world" is just waiting for us to screw up and prove them right.

Many of us deal with such insecurity in our own abilities that we latch on to someone who seems to have at least most of the answers. I'm speaking from experience here. I spent years thinking that if I could just hook my wagon to a "shooting star" curriculum or educational method, I would be safe — I would be guaranteed a decent outcome. Thus, I spent years being dedicated to the cause of not failing. I allowed the fear of failure to make my decisions, I allowed the pursuit of validation to be my focus, and I let my insecurity set the tone of my life and the atmosphere of my home.

I have to be honest with you; in recent years, I've become increasingly concerned about how easy it is for us to use social media in this humanly natural pursuit of validation. I'm going to tread lightly here… I don't want to make social media out to be a villain, because the issue truly is a heart thing. Therefore, I do want to boldly challenge all of us to allow God to examine our hearts. Do we spend time on social media platforms in either of these two ways?

Number one: Have we turned our social media into a reality TV show, starring us? Do we live differently than if there were no cameras, no "go live" button, and no way for us to show the world how cool we are? Are we using it as a way to find significance and validation?

Number two: Do we go to certain people on social media for reassurance that we are good enough? Do we go "bask in the glow" of someone we think has it all together and then try to be like them?

I know that these are hard questions, and the very nature of social media is, in many ways, an invitation to congregate on the "platform of fake." I want to challenge all of us to stay authentic and plugged into our real lives with the real people God has given us to disciple. Our entire friends list doesn't need to give us a constant thumbs up about our lives; we are here to please God.

I'm going to make a bold statement because I have made a covenant with God to remain in this position: I do not write, speak, or advise to gain a following. I want you to be my friend and fellow homeschool parent, not my "follower." I want us to be fellow followers of Christ, brothers and sisters, children of the Almighty God, and co-workers in the Kingdom work to which God has assigned each of us.

In fact, it is my goal to consistently work myself out of a job. If I encourage you to continually come back to sit at my feet for advice and tips, I'm not doing you any service at all. I want to give you the kind of help, tips, and encouragement that sends you on your way, into your own life, with Biblical confidence that you can do this, because God has called you to do it. And He, not I, will give you the strength and wisdom that you need to face the challenges in your life. If my words do not lead you to Him, then I do not want you to read them. I am human, and even if I try really hard, I will let you down, because I am fallible. I do not have all the answers to your problems. What I do have, I give to you freely, but your heavenly Father is the only source of truth, peace, divine wisdom, and true joy. Let's promise each other that we will go to Him together and walk through life in His strength.

Always remember that you are being divinely equipped to educate your children. As a homeschool mom myself, I know how important it is to remember on a daily basis that we are all being homeschooled by God. He is the One who created your family, and He is proud of your obedience and your effort. You are not alone; He sees you, He knows you, He loves you, and He is with you. He did not call you to do this and then leave you to figure it out on your own and complete the assignment in your own strength. You are part of a much bigger story than you could ever imagine. What you do matters. All those holes left by our human weakness are a perfect showcase for God's grace and glory. May we all face each day with a heartfelt *Soli Deo Gloria.*

Imitate God, therefore, in everything you do,
because you are his dear children.
Live a life filled with love,
following the example of Christ.
He loved us and offered himself as a sacrifice for us,
a pleasing aroma to God (Ephesians 5:1-2; NLT).

Note: You can supplement the worksheets in the *Math for a Living Education* series with additional worksheets, activities, and quizzes in Practice Makes Perfect, also available from Master Books.

Part 1

Understanding the Method and Approach of *Math Lessons for a Living Education*

Section #1: The Biblical Worldview Method in Math:

We, as a culture, have been persuaded into thinking that only the professionals are able to educate and socialize our children. As parents who have chosen to home educate, we know that this is simply not true. The more I've studied this phenomenon, the more I have come to realize that much of it is rooted in the secular worldview that discounts God and the divine sanctity and strength of family relationships. We have been taught that teaching our children academic subjects requires a separate sphere of intelligence and ability than teaching them how to speak, walk, and go potty; that this ability is not found in the common, untrained parent — the category into which most of us fall. Because of this, even in the homeschool world, there is a large amount of curriculum with classroom-style, overly-complicated, scripted "teaching instructions" that has allowed the "experts" to teach our children through us, instead of trusting the Holy Spirit to lead us and give us wisdom for guiding our specific children in our own unique home education setting. (James 1:5)

We are led to believe that teaching elementary math and language concepts is in the same league as rocket science, even though we all use these concepts on a daily basis without even thinking twice about it. We are told that our children, regardless of their uniqueness, can only learn through certain types of activities, which follow mysterious scopes and sequences (known only to the "professionals"), at a specific time in their lives. Thus, we the parents — the very ones that God chose before the creation of the world to be the nurturers, disciplers, shepherds, and teachers of our children's minds, hearts, and souls — shuffle through the cold cloud of insecurity and fear that we are going to mess our kids up, let them down, and cause them to be behind some arbitrary standard. This is the cultural stronghold

we homeschooling parents face on a daily basis, and we need to constantly and consistently remind ourselves and each other that we have truth on our side.

Let me assure you: nowhere in God's Word does it mention that children should be taught by someone who cares little to nothing about their spiritual development or eternal purpose. Nowhere is it written that every child needs to perform and be measured by manmade schedules, standards, and scope and sequences. On the contrary, in the Bible, all instructions about children are built first and foremost upon their spiritual development as the foundation. And these directives are primarily aimed at the parents (with an occasional mention of grandparents). As homeschooling parents, we understand this; in many cases, this is the underlying reason for our decision. However, most of us have a hard time breaking loose from the cultural mold that has trained us to believe that we are not capable of implementing it in our own homes, and, at the same time, produce children who are equipped to live in our current culture.

In the ancient days when God called His chosen people, He gave them clear instructions on how to raise and train their children with an education centered on their spiritual development. These directives are still extremely pertinent to us today. You are probably familiar with them — they are called the *Shema*, and we can find them in Deuteronomy 6:4-9. The *Shema* is the divine prescription to us as godly parents to raise our children to be healthy, grounded adults who have a solid view of who God is, who they are, and how to interact with the world around them.

I boldly challenge us all to believe this. If we follow it, the truth in these Scriptures will replace the lies about what the world calls "education," and we **will** raise children who are ready to stand strong in this current day's culture. They will be in the world but not of it. This is the Hebrew model (not method) of education; it is centered around reaching the heart of the child with the goal of training them — body, mind, and spirit — in a godly, biblical worldview in every subject, including math. We are called to be disciples of Christ — students of His ways, His Word, and His Gospel. As parents, we are required to model this for our children.

I believe that a good curriculum teaches the whole child, but an excellent curriculum is a discipling tool that guides, reassures, and supports the parent and reaches the heart of the child, pointing the whole family to Christ. This is the foundational goal for the *Math Lessons for a Living Education* series. The entire series is meant to be a tool for you to help you raise, train, and teach your children in the way they should go.

This may sound overly dramatic, but in many ways, I feel like God has called me to work (alongside several others, including Katherine Loop Hannon,

author of the *Principles of Mathematics* series) in one of the last frontiers of home education: *Math*. For decades, homeschooling families have been aware of the worldview presented in every other subject. They have been teaching apologetics in science, God's hand in history, and glorifying Him through reading and writing. And there, on that lonely shelf, very often surrounded by contempt, is math. It does not have to be this way.

Teaching and learning math isn't only about number concepts; no subject is ever just about gathering information into the brain. It's about bringing glory to God in every area of life. It's about seeing Him as an integral part of every aspect of the universe.

Being educated is not a distant goal that we are striving for; it's in the process of learning to learn — the growing and becoming who God created us to be, day after day — that we become truly educated. This process is not something that humans can regulate or control with their lists and standards, it is encoded in our very being — our made-in-the-image-of-God DNA.

Section #2: "Good-brain"* Math

Many of us grew up with math curriculums that made us mad. Quite honestly, I disliked math as a small child — I should say, I disliked the math curriculum that I was made to do in those years. The endless pages of repetitive equations seemed like they were designed to drive me out of my mind. I remember feeling like whoever wrote those books strongly disliked children (and had never been one themselves) and had designed the lessons in this way in order to trick the tired student into making a mistake. I truly was convinced that it was a rigged system. This was "bad-brain" math. It sent me into a negative thought cycle. I did not truly understand this phenomenon until I was older and studying the effect that negative thinking has on our learning ability and experience.

Since the death of my own dad, with whom I shared a special bond, I have had the privilege of

being mentored by several very godly men who have become like my spiritual fathers. One in particular has helped me to understand the importance of being a "good-brain" person in my children's lives. This mentor, Professor Gary Newton, is a pastor, a professor of Christian family life and discipleship ministries, an author, speaker, and profoundly insightful life and leadership coach. Gary's book, *Heart-deep Teaching*, made a powerful impact on me, swinging my attention away from my own agendas and fear of failure and towards creating a learning environment that facilitated a positive, "good-brain" educational experience.

Dr. Newton and I had many long discussions while I was creating the original editions of *Math Lessons for a Living Education*. In one of the most profound conversations I had with him, Dr. Newton told me that the most successful and outstanding students

---

* The term "good-brain" is borrowed from my friend, Randy Pratt of Master Books.

in his college classrooms are the ones who have been encouraged to make a relationship with their education — in *every* subject. These are the ones that had been taught to approach their education with determination to develop their critical and creative thinking abilities.

He also told me that he could tell which students had spent their lives filling in the blanks and being taught to the test. These students struggled to think outside the box; they did not do well in figuring out how to deal with unexpected disturbances in life and tended to be followers instead of leaders. They were stuck in what I call the "bad-brain" learning mode, with underdeveloped brains.

I believe firmly that we are all called to be "good-brain" people in our children's lives…encouragers, allies, and high-standard-holders. Our kids need to know that the potential for greatness is in them because they are created in the image of God. They need to learn to see their education as an exciting adventure — a private journey between them and their Creator. We need to instill the truth of the greatness and accessibility of God so they don't doubt that He is waiting for them to come to Him so He can show them the good plan He has for their lives. We, as parents are the guides for

them on this journey. I believe that the more we build our personal relationships with God first, the relationships with our children that show them the way to God will naturally follow. Luke 6:40 (NKJV) says, "A disciple is not above his teacher, but everyone who is perfectly trained will be like his teacher." We cannot expect our children to be something that we are not modeling for them.

When I wrote *Math Lessons for a Living Education*, one of my central goals was to create "good-brain" math. I wanted parents who had lived through a "bad-brain" math experience in their own childhood to be able to learn alongside their child with no judgment or condemnation that they "should already know it" or that "they stink at math." I believe with all of my heart that if we can keep ourselves and our kids out of the "bad-brain" approach to learning, our educational journey will be substantially easier and more successful.

Is *Math Lessons for a Living Education* different than anything else you've seen or experienced?

I sure hope so.

Section #3: Goals and Objectives

My goals for *Math Lessons for a Living Education* are simple and straightforward:

- To demystify teaching and learning elementary math, to nurture the love of learning in the child, and to encourage the confidence to teach in the parent(s)

- To give children a firm foundation in math concepts that will serve them for the rest of their lives

- To give families a tool for learning and growing together at the speed and in the style that fits their needs

- To encourage the child to connect with their learning journey and help them to become life-long learners

- To be a good-brain person in your child's life and in your life… cheering you on and pointing you to Jesus

Section #4: Scope and Sequence Chart

Notes about *Math Level K* are included after the chart for Math Levels 1 - 6.

Lesson #	Math Level 1	Math Level 2	Math Level 3	Math Level 4	Math Level 5	Math Level 6**
1	Numbers 0 - 9	*Place Value Village, Telling Time, Shapes & Patterns	*Review of Place Value, Odds and Evens, Counting by 2, 5, and 10	*Review of all Addition and Subtraction Concepts	*Review of all Addition and Subtraction Concepts	Working with Whole Numbers
2	Numbers 0 - 9	*Addition, Horizontal & Vertical, Shapes	*Review of Money, Clocks, Perimeter, Addition/Subtraction Facts	*Review of Place Value, Estimation, and Rounding	*Review of all Division and Multiplication	Whole numbers in the Real World
3	Introducing Rectangles	*Subtraction	*Review of Addition, Including Carrying, Tally Marks	*Review of all Multiplication	*Review of all Geometry	Averaging, Rounding, and Roman Numerals
4	Circles and Patterns	*Writing Numbers to 100, Simple Fractions	*Review of Subtraction, Including Borrowing Concepts	*Review of all Division	*Review of all Measurements	Fractions
5	Review of Concepts	Introducing Word Problems	*Review of Measurement, Fractions, Thermometers/ Graphs	*Review of all Fractions and Measurement	Review of all Fractions Concepts	Working with Factors
6	More Numbers, Patterns, Shapes - Introducing Triangles	Skip Counting Using Dimes and Nickels, Minutes on the Clock	*Review of Word Problems	*Review of all Roman Numerals and Shapes	Review of all Decimal Concepts	More about Fractions / Mixed Numbers
7	Bigger Numbers/ Place Value	Skip Counting by 2, Even and Odd Numbers	Introducing Column Addition/Larger Numbers	Fraction Concepts: Adding and Subtracting Like Denominators	Multiplying and Dividing by 10, 100, 1,000	Using Factors and Multiples in Operations
8	More Work with Place Value	Addition: Double Digit Plus Single Digit	Introducing Larger Number Subtraction	Multiplication with Carrying Using 11s and 12s	Introducing 2-Digit Divisors	Review of Fraction Concepts

* Indicates review lessons - concepts in previous levels. Additional optional teaching instruction included in this *Teaching Companion*.

** *Math Level 6* follows a slightly different design than *Levels 1-5*. *Levels 1-6* have 180 days' worth of work, while *Math Level K* is set up on a 3 day/week schedule, with 2 optional days to create an alternative 180-day schedule. The Lessons in *Math 6* cover more than five days in the schedule to create 180 days.

Math Teaching Companion

Lesson #	Math Level 1	Math Level 2	Math Level 3	Math Level 4	Math Level 5	Math Level 6**
9	Review of Concepts	Addition - Double Digit Plus Double Digit	Introducing Rounding to the 10s and 100s	New: Measurements and Geometric Concepts	More Work with Division	Adding and Subtracting Fractions and Mixed Numbers
10	Place Value/ Patterns of 10s	Review of all Addition Concepts	Adding and Subtracting Larger Amounts of Money	Review of all New Concepts	Three ways of Division/ Remainders as Fractions	Multiplying and Dividing Fractions
11	Practice with Patterns and Shapes	Introducing Measurement/ Inches, Feet, Review Time & Shapes	Review of all New Concepts	Steps of Division/Single Digit Divisor, No Remainder	Review Week	Comprehensive Review of all Advanced Fractional Concepts
12	Introducing the + and = Symbols	Introducing Perimeter	Introducing Multiplication of 0, 1, 2, and 5	Number Grouping/ Understanding Larger Multiplication	Factoring	Decimal Basics
13	Addition +1	Telling Time to the Minute	Introducing Division of 1, 2, and 5	More about Division Including Checking Division	Common Factors, Greatest Common Factor, Reducing Fractions	More Work with Decimals
14	Writing and Adding Numbers/ Intro: Days of the Week	Place Value Village Practice/ Place Value to the Thousands' Place	Introducing Multiplication and Division of 10	Division with a Remainder (Single Digit Divisor)	Proper and Improper Fractions	Using Decimals in the Real World
15	Vertical Addition	More Work with Subtraction	Introducing Areas of Rectangles and Squares	Metric Unit of Measure	Working with Improper Fractions	Percents
16	Introducing Squares	Introducing Addition with Carrying to the Tens' Place	Introducing Multiplying and Dividing by 3	Review of all New Concepts	Sums Containing Improper Fractions	Using Decimals and Percents in the Real World / Savvy shopping

* Indicates review lessons - concepts in previous levels. Additional optional teaching instruction included in this *Teaching Companion*.

** *Math Level 6* follows a slightly different design than *Levels 1-5*. *Levels 1-6* have 180 days' worth of work, while *Math Level K* is set up on a 3 day/week schedule, with 2 optional days to create an alternative 180-day schedule. The Lessons in *Math 6* cover more than five days in the schedule to create 180 days.

Lesson #	Math Level 1	Math Level 2	Math Level 3	Math Level 4	Math Level 5	Math Level 6**
17	Two by Two / Skip Counting by 2s	Introducing Subtraction with Borrowing from the Tens' Place	Taking Fractions Deeper	Introducing Mixed Numbers / Adding and Subtracting with Like Denominators	Least Common Multiples	Comprehensive Review of Fractions, Decimals, and Percents
18	Number Families, Addition to 10	Review of Regrouping Concepts	Multiplying and Dividing by 4	Introducing Equivalent Fractions Through Pictures	Least Common Multiples/ Finding a Common Denominator	Geometry
19	Counting by 10	Understanding Dollars and Cents, Writing Money Terms	Multiplying and Dividing by 6 and 7	More About Equivalent Fractions	Review of all New Concepts	Maps! Just Follow the Lines
20	Counting Groups	Review of Money	Multiplying and Dividing by 8 and 9	Larger Number Multiplication with Carrying	Adding Fractions with Uncommon Denominators	Graphs and Charts
21	Solving for an Unknown	Introducing Thermometers and Other Gauges	Review of all New Concepts	Review of all New Concepts	Subtracting Fractions with Uncommon Denominators	Units of Measurement
22	Tally Marks to Make Groups of 5	Reading Bar Graphs and Line Graphs	Rounding to 1000s and Estimation	Writing Decimals and Fractions	Subtracting Mixed Numbers with Carrying & Borrowing: Common Denominators	Additional Topics
23	Counting by 5s	More on Measurement: Pounds, and Ounces	Higher Place Value Through Millions	Money Work with Decimals and Fractions	Adding Mixed Numbers with Carrying: Uncommon Denominators	
24	Telling Time #1	More Measurement Concepts: Gallons, Quarts, Pints, Cups	More Measuring Concepts	Relationship Between Fractions, Decimals, and Percents	Subtracting Mixed Numbers with Borrowing: Uncommon Denominators	
25	Telling Time #2	Review of Measurements	Introducing Solving for Unknowns	Geometry	Review!	

* Indicates review lessons - concepts in previous levels. Additional optional teaching instruction included in this *Teaching Companion*.

** *Math Level 6* follows a slightly different design than *Levels 1-5*. *Levels 1-6* have 180 days' worth of work, while *Math Level K* is set up on a 3 day/week schedule, with 2 optional days to create an alternative 180-day schedule. The Lessons in *Math 6* cover more than five days in the schedule to create 180 days.

Lesson #	Math Level 1	Math Level 2	Math Level 3	Math Level 4	Math Level 5	Math Level 6**
26	Telling Time #3	Adding Money - No Regrouping	Introducing Inequalities	More Geometry	Multiplying Fractions	
27	Introducing Simple Fractions #1	Subtracting Money: Making Change	Review of all New Concepts	Review of all New Concepts	Divisibility Rules and Dividing Fractions	
28	Introducing Simple Fractions #2	More Work with Word Problems	Addition and Subtraction of Larger Numbers	Work with Charts and Graphs	Multiplying Decimals	
29	Introduction to Subtraction	More Work with Telling Time	Introducing Roman Numerals	Constructing Charts and Graphs	Hands-on! Counting Back Money	
30	Subtraction -1	More Work with Measurements	More About Roman Numerals	Introducing Averaging	Review of all Division	
31	Review of Shapes	Review of Place Value Through the Thousands' Place	Review of all Addition and Subtraction Concepts	Review of all Addition and Subtraction	Review of all Division	
32	Review of Place Value: to 100	Review of Word Problems: the Steps of Solving	Review of Rounding, Estimation and Place Value	Review of all Multiplication and Division	Review of Factoring, Common Factors, and Greatest Common Factors	
33	Review of Addition	Review of Adding and Subtracting: Double-Digit Problems	Review of Multiplication	Review of all Geometry	Review of Fractional Concepts #1	
34	Review of Skip Counting, 2s, and 5s	Review of Money Concepts	Review of Division	Review of all Measurement	Review of Fractional Concepts #2	
35	Review of Skip Counting 10s and Tally Marks	Review of Time and Temperature	Review of all Measurements, Fractions	Review of all Fractional Concepts	Review of Multiplying and Dividing Fractions	

* Indicates review lessons - concepts in previous levels. Additional optional teaching instruction included in this *Teaching Companion*.

** *Math Level 6* follows a slightly different design than *Levels 1-5*. *Levels 1-6* have 180 days' worth of work, while *Math Level K* is set up on a 3 day/week schedule, with 2 optional days to create an alternative 180-day schedule. The Lessons in *Math 6* cover more than five days in the schedule to create 180 days.

Lesson #	Math Level 1	Math Level 2	Math Level 3	Math Level 4	Math Level 5	Math Level 6**
36	Review of Numbers to 100	Review of Addition and Subtraction Fact Families	Review of all Roman Numerals and Shapes	Review of all Decimal Concepts	Review of Multiplying and Dividing Decimals	

* Indicates review lessons - concepts in previous levels. Additional optional teaching instruction included in this *Teaching Companion*.

** *Math Level 6* follows a slightly different design than *Levels 1-5*. *Levels 1-6* have 180 days' worth of work, while *Math Level K* is set up on a 3 day/week schedule, with 2 optional days to create an alternative 180-day schedule. The Lessons in *Math 6* cover more than five days in the schedule to create 180 days.

Math Level K:

The focus of *Math Level K* is early childhood development that is needed for *Math Level 1* and up. Many children in this age bracket are almost cognitively ready to begin learning the actual number concepts but need a little extra time to develop the motor skills necessary to perform many of the tasks included in *Math Level 1*. For these young ones, it is crucial that they have the opportunity to build the necessary prerequisite skills before they begin their acquisition of the more advanced small motor skill activities. You will find that *Math Level K* helps the young child to learn skills such as understanding directional concepts (left, right, first, last, next, etc.), recognition of basic shapes and colors, one-to-one matching, and beginning numeric recognition and understanding.

Through the well-loved stories (including the long-asked-for prequel of the story of Charlie and Charlotte!) and hands-on activities *Math Lessons for a Living Education* is known and loved for, young children will learn by leaps and bounds and be well-prepared for the next step of *Math Level 1*. *Math Level K* is written to coordinate with the new Master Books Kindergarten curriculum, *Simply K - a Developmental Approach to Kindergarten* by Carrie Bailey (who is also the co-author of *Math Level K*). By using both *Math Level K* and *Simply K*, you can give your child an enjoyable and thorough foundation on which to build their elementary education.

Part 2

Teaching *Math Lessons for a Living Education*

Section #1: Making *Math Lessons for a Living Education* Work for You

Using *MLFLE* as a Tool

Math Lessons for a Living Education is an out-of-the-box curriculum, which, instead of forcing the child to complete pages of drills, asks them to develop the communication between their creative and critical thinking skills in brain-growing ways. Because of this unusual approach, I am often asked if it is enough to be a complete curriculum on its own. The answer is a resounding, "yes." Elementary students do not need to spend an hour "doing" (our school word for writing and memorizing formulas) math problems; they need to actually have time to play with the concepts that they are learning. Children need to play to learn; in fact, it's been said that playing is a child's work. Children need to be allowed to take the concepts that they have heard and seen and be able to internalize them through play. As they get older, children turn their play into other types of creative thinking. Thus, play is the foundation of both creative and critical thinking.

Now, of course, this doesn't mean that your child needs to have their seatwork time, working on their math lesson, plus a structured half-hour of "playing to internalize the concept" time. It simply means that we, as the parent-teacher, need to not take up all of their time with structured instruction. Charlotte Mason called this practice "masterly inactivity." This approach of having a balance between involvement (the teaching and discussing) and allowing for the development of thought, allows the teacher to encourage the child to own their education at a very young age. You will see that in *Math Lessons for a Living Education*, I have included many interrelated projects to bring a playful, hands-on aspect to the program and to help the parent and student to transition into more of an out-of-the-box way of thinking.

You are going to see and hear me say the words "the communication between creative and critical thinking" many times in this *Companion*. This is because, contrary to what you might have been told, creative and critical thinking walk hand-in-hand; one is not complete and competent without the other. Although they may come from different physical parts of your brain, they are absolutely interwoven and intertwined in practice. You simply cannot have well developed critical thinking skills without having well developed creative thinking skills, and vice versa. This is perhaps evidenced in the study of math more than any other discipline. For example: think about solving a word problem. Many people do not like word problems because they feel lost in the information. Their left brain says, "Way too much fuzzy information going on here," and their right brain says, "Not happening! Far too many numbers and too much logical thinking required!"

Anyone can learn to become efficient at solving word problems by developing the ability to mentally see the situation described by turning it into a picture (right side work). They can then organize the numeric information being given into a logical mental display (left side work). Next, they can ask themselves, "What is the question I'm trying to answer? What is the missing information?" (left side work). After determining the question, they can then choose the operation(s) needed to find that answer (left side work). Finally, they can use their pictured, organized information to plug in those operations to find the missing information — the answer to the word problem (left and right side working together).

If we do this often enough, we can train our brains to approach word problems like a puzzle. This process takes both creative thinking and critical thinking working together. Most people are not taught to approach math with a creative mind, but the truth of the matter is this: our brains were not meant to be used one side at a time. The left side may be analytical, but the right side has to do the expressing of that analytical thought. We need to help ourselves and our children to train our brain hemispheres to work together.

Adapting for Cognitive Ability and for Your Unique Child

First and foremost, ask for divine insight and wisdom to see what you need to do to best adapt to your child's ability and learning style, and then learn to trust that insight and wisdom. (James 1:5)

Math Lessons for a Living Education was written in such a way to be extremely adaptable to the individual student's ability. If your student is grasping the concepts at a faster rate than the lessons are laid out, please allow them to advance at their own speed. Chances are that they will slow down at some point when they encounter a concept that takes them a bit longer. On the other hand, don't fall into the trap of comparison and worry. I am learning that both of these are fueled by fear and pride, neither of which are from God. (Proverbs 16:18 - Pride)(2 Timothy 1:7 - Fear)

For special needs students, please use this curriculum as a tool in whatever way you need. I suggest reading the stories together and working through the problems together. Work as slowly as your student needs. I have known many special needs children who took 8 to 10 years to complete the entire scope of the math series. By the time they were finished, they were able to learn how to keep a checkbook and take care of their own finances. Don't forget, the concepts covered in these books are the most commonly used math concepts we all use in everyday life.

For advanced students, my only warning is to make absolutely sure that your student can narrate to you the steps of the concept and why they are doing them. Many students are deemed "good at math" because they have become proficient in "monkeying" and filling in the blanks. The success of actually being able to understand math in real life is based on being able to understand the when and why behind it. I have seen numerous cases where a mother will say, "My child gets it but doesn't like to narrate the process of why they are doing it, so I just let them fill in the blanks and do the work in their book."

Please Understand:

- Your student, no matter their learning style, needs to be able to articulate their thoughts. In the discipline of writing, they need to be able to articulate in complete sentences. In math, they need to be able to articulate why they are doing what they are doing. This is not just about writing or math, this is about logical and critical thinking and being able to process through the steps of being able to output what is in their brains. I do understand that this can be a painful process and therefore have given you tips in Part 2, Section 5 for helping them through this process and understanding oral narration in math. (Please read the note at the end of the previous section about critical and creative thinking.)

- Although *Math Lessons for a Living Education*

is not written to meet an arbitrary grade level guideline, because there are seven volumes, it makes sense that they will fall approximately in the same age range as all other math curriculums written to be specific grade levels. However, *please* do not approach this math series like all others. Allow your student's ability to show you where to place them. If entering this series from another curriculum, please give your student the readiness test which is available in the back of this book. You can also download the test at MasterBooks.com.

Adapting *Math Lessons for a Living Education* for Learning Styles

Math Lessons for a Living Education was crafted to include each of the learning styles.

- **For visual learners**, the bright colors, graphics, and pictures will be helpful in their learning journey, as will the plentiful amount of white space on the pages. There is also the probability that visual learners will want to read the story to themselves. It has been my experience that visual learners become independent a bit faster than other types of learners. Visual learners tend to have high visual comprehension levels that lend themselves to being able to read and understand instructions and story problems (of course, this may or may not apply to your visual learner). Make sure that you are connecting with your visual learner to make sure that they can verbally explain what they are doing. Visual learners tend to be good at fill-in-the-blank. You may also find that your visual learners move more quickly through the levels of learning, from the need for hands-on manipulatives to the ability to work problems in their heads. Many times, visual learners have a harder time than auditory learners in expressing and articulating the concepts and processes they have learned.

- **Most auditory learners** need to hear the story, directions, and story problems. Two of my kids are auditory learners, with my youngest child being an extreme case. She is so auditory that we have to work hard to build visual comprehension. Auditory learners may want to read the story, instructions, and story problems aloud. I allow my young auditory learners to read everything out loud, or I read it to them. Also, auditory learners tend to need discussion

and interaction more than visual learners do. They need to be able to talk through the concepts and discuss each of the steps. Auditory learners are generally more naturally fluent in their oral narration abilities.

- **Kinesthetic learners** retain best by getting actively involved in their learning process. A simple way to turn any math concepts and lessons into a more active learning experience is to allow your student to change the way they approach their lessons. For example, sitting on an exercise ball or standing up while doing their math work can change how the brain functions. This is why I suggest that your child use a whiteboard to do some of their work. Many kinesthetic learners have a hard time sitting still during their studies and so are often deemed "wigglers" and "fidgeters." These kiddos truly need to "move and shake" in order for their brains to take in and process their sensory input. Several of my kids benefitted from "thinking putty" that kept their hands busy while they were thinking.

- Of course, most of us are a mixture of at least two of the learning styles. If you are not sure what your child's learning style is, talk to them about it. Pay attention to what they are doing when they are concentrating. Take notes about their need to either see or hear the instructions. No matter what your child's learning style,

always remember that we all retain best while using as many senses as possible. Note: many times children who are primarily auditory learners are also kinesthetic. They benefit greatly from discussion and interaction. On the other hand, it is quite common for visual learners to be better memorizers.

- Add hands-on, out-of-the-box practice, and application. In Part 4 of this *Companion*, I have included a list of ideas for learning math through games and other hands-on projects. For additional tips and ideas on teaching, please read the next section, General Teaching Tips.

General Teaching Tips

- Always start with a review. We all do much better when we start with something familiar and build on it. This can be done simply by quickly reviewing the previous lesson's concepts before moving to the new lesson's content.

- Engage your student in the lesson. Invite them to join in on reading the story, discussing the instructions, playing with the manipulatives, and memorizing facts. Interaction is the key to learning.

- Encourage your child to ask questions. You can't fill or pour from a pitcher that has a closed lid, and a child can't learn with their mouths shut. From a very early age, encourage your child to ask questions. Help them learn to ask good questions that have meaningful answers. If you don't know the answer, learn to research with your child to find the answers. If your child is older, start now. If your child is a toddler, start now.

- Keep a small container or an expand-a-file with several colors of index cards on hand. As you and your child are working through the concepts, work together to make "study cards" showing the breakdown of the concepts. I've even had my students demonstrate the concepts with manipulatives, while I took pictures of the process. Then, I printed the pictures and pasted them to the study cards. I have done this in every subject. A couple of my kids ended up with two large shoe boxes of study cards by the time they graduated.

- Allow God to help you deal with any anxiety that you may be experiencing. Your relationship with your Heavenly Father is the most significant deciding factor in your success as a homeschooler. Actively be establishing your own biblical God-view, worldview, and self-view. Be more concerned with your own diligence in seeking His daily direction than in keeping a schedule for your children's schoolwork. "But seek first the kingdom of God and His righteousness, and all these things shall be added to you" (Matthew 6:33 NKJV). I promise, if you seek His direction, He will lead you on paths of peace.

- Pray for guidance and then trust that guidance. Put blinders and earplugs on if you need to keep from being distracted from that path (speaking from experience).

I have created *Math Lessons for a Living Education* in such a way that allows you as the homeschool parent to shine. My intentions for keeping lessons short and giving direction in becoming strong in critical and creative thinking are to allow you to have time to do what you can do best: guide your child in the way they should go, with God's help.

I highly encourage all parents, even ones who have been on this journey for a while, to schedule an afternoon or entire day — depending on how many children you have — to sit and think about each child. Give each one a whole page or two in your thinking journal (you could benefit greatly to have one of these if you don't already — a simple, single-subject spiral works great) and write down these questions to answer:

- What are our spiritual goals for this child?
- What are our character development goals for this child?
- What are our relationship goals for this child?
- What are our academic goals for this child?
- What are other areas where we need to be seeking divine wisdom concerning this child?
- What is the Scripture we will be praying over this child's life?

I am discovering that it is a lot harder to get deeply upset, angry, and pessimistic about someone for whom I am fervently and consistently praying. Frustration with a person can much more easily be replaced with beneficial thoughts and supplications when you have a practical and spiritual plan of action for that person.

Section #2: Teaching Place Value with Place Value Village

Constructing the Place Value Village (PVV)

Place Value Village is included in the Manipulatives Section of *Math Level 1-3*. To construct the village, simply cut out the houses, allow your child to color them (optional), laminate* them for sturdiness (optional), and attach them to some type of container. You can use any kind of container you want

that will hold the counting items. I've seen moms use everything from stacking storage containers to large plastic drinking glasses. If you are wondering what the finished product should look like, follow the link below to see the tutorial I made for PVV. I do recommend having a basket or bin that is specifically for storing your Place Value Village, so you can keep all of the pieces together.

Using Counting Objects

Any type of small item can be used for teaching the Place Value Village. In the books, I suggest using dried beans because they are affordable, and they are easy for little fingers to handle; however, you can use any small item you want. In our family, when my older kids were young we used the beans, but my younger two were gifted a set of one thousand linking math cubes. Both the beans and the cubes worked great to help my kids understand place value. There isn't a specific small counting item that is going to help or hinder the learning process. Please feel free to use any small items you have in your house already or, if you would prefer, go buy some new ones.

Understanding the Learning Process of This Concept

Children will not completely understand place value the first time they are exposed to it. It is a concept that needs to be visited often over time.

---

* I have never owned a laminator, but I have used clear Contact Paper ™ to cover school projects like this. I've even seen some moms use clear page protectors. You can laminate the PVV houses in any way you want. There is no right or wrong way to do this.

Don't panic; this is normal and acceptable. You will probably discover that your child "gets it" sometimes, but not other times. This is also completely normal (human brains are funny like that - they work in 'spits and spurts'). Simply understand that you, as the teacher, are going to be working with and adding onto this concept very frequently while your child moves through the concepts covered in the *Math Level 1-6* levels.

There is a tutorial video for the Place Value Village at this link: https://angelaodellblog.com/2016/04/11/place-value-village/

If your student is balking over learning of this concept, explain to them that they are building a skill that takes a while to fully construct in the mind. It is most certainly not a one-time-and-you-have-it type of concept (actually, very few things in life are!). As their cognitive ability grows, so will their understanding of this concept. They are simply encouraged to do their best and practice often. Note: place value is certainly one of those concepts that requires critical and creative thinking to work together.

Likewise, if they do understand it at its simplest level, don't expect them to continue to understand it perfectly as they move up into the more difficult levels of the concept.

There will probably be times that they will have to slow down, go back, and reinforce the simpler levels before they can move on to continue building on it. It's perfectly fine if this happens. Needing to review is not a sign of failure on anyone's part — it is simply something that is required in the learning journey. Simply keep reviewing and moving forward.

Set aside planned days to focus on building place value understanding. Ideas: set up the Place Value Village on the table, use fun (and unusual) counting items, make a special treat to snack on, put some fun music on, etc. The idea here is to create a fun, festive atmosphere that is not ordinary.

Think of it this way: when you, as an adult, are learning a new skill or concept, you are constantly subconsciously thinking about what you know and then slowly adding to that base. Our kids are just finding out how they don't know everything, how to understand the world around them, how to build a little confidence through incremental growth and learning, and how to have a positive attitude about the learning process. They are not only learning the concept of a number, but they are also training their brain and building their character.

I have included hands-on activities in the Math Games section in Part 4 of this *Companion* to use in addition to the Place Value Village.

I have included a chart for you to use to take notes about your child's journey through the concept of place value. Because every child is different and will travel at their own rate, and because you know your child best, I want you to simply keep track of their journey. Remember, this is not about a goal of reaching comprehension at a certain time. This is about journaling about the journey. Relax and enjoy the trip.

Child's name	Notes About My Child's Learning Journey: Place Value		

We know as adults that the learning process can be fun, but it can also be humbling. We have to admit when we don't know how to do something and be teachable to learn it. This concept is huge for all of us, as it translates into the emotional and spiritual parts of our lives as well as the educational.

Section #3: Right-brain Flashcards

The Method and Objectives:

Right-brain flashcards are yet another creative-critical-thinking team player exercise. Honestly, a lot of kids (and parents) have a hard time with them. Because they are so "unusual and different" from other approaches, they cause an uncomfortable feeling or even plain old insecurity. I remember feeling all of the above. Once you understand why you or your child are feeling this way, it is much easier to address it and deal with it. It makes us uncomfortable because our brain's analytical left side doesn't want to have to communicate with the creative right side; it just wants to do its thing, fill in the answer and move on. Many times, the left side is not good at communicating and is even a little possessive of the more analytical and critical skills. According to our brain's left side, the right side is "hairbrained" or maybe even a lesser twin. The process might be painful at first, but it is oh, so beneficial. When we train our brain hemispheres to communicate, we are creating pathways between the two, which in turn strengthens the ability to reason and communicate fluidly in all areas of life, raising the level of our verbal aptitude and cognitive ability. Imagine that you are helping your child to train their brain hemispheres to walk together, each taking its turn instead of one side dragging the other or trying to exclude it completely. This is a discipline worth pursuing.

Right-brain flashcards help the student to memorize the whole fact by not allowing the student to see the equation with a blank for its answer. This is especially important for visual learners. (This is similar in concept to not allowing a child to see a word misspelled if at all possible. It is much easier to start with the correct spelling and the whole fact than to correct a wrong answer habit later.)

Right-brain flashcards do not have to be fancy. They can simply be the math fact, including the answer. You do not have to be super creative to make this type of study aid. If your child wants to make up a story for the fact they can, but this is entirely optional.

Right-brain flashcards can be horizontal, vertical or (for fact families) triangular. Study the samples of the addition/subtraction flashcards and multiplication/division flashcards below.

Please note: Subtraction is introduced in Lesson 29 of Math Level 1. Therefore, in *Math Level 1*, do not use the Triangle Flashcards for your right-brain flashcards.

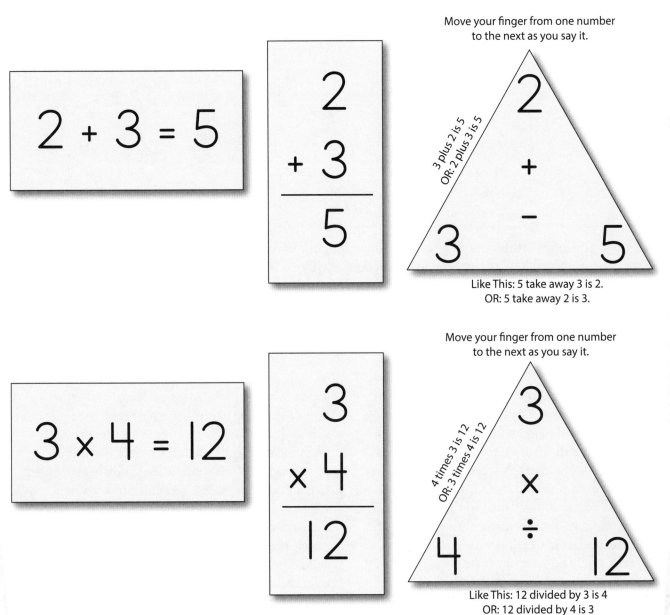

Creating the Cards

What you will need to create your right-brain flashcards:

☐ index cards

☐ card stock cut into triangles with 3 or 4-inch sides

☐ pen or pencil

Optional:

☐ stickers

☐ markers

☐ any other craft items that can be used to "jazz up" your creations

How to make them:

1. Write the entire fact on the card.
2. Have your child practice often.
3. Optional: decorate the card and make a story to match the fact.

Section #4: Memory Work and Math Facts

The Facts about Math Facts

- **You cannot force someone to memorize something;** you can work with them to find what will help them learn. Please don't turn it into a battle.

- **Memorizing math facts makes life easier.** Sometimes when pokey, unmotivated memorizers hit the place in their math journey where they run out of fingers and toes to count on, they see the value of memorizing and therefore begin to invest in it. Sometimes not. You still can't force someone to memorize, but you can pray for wisdom to understand the underlying issues.

- **Generally speaking, when a child refuses to memorize math facts (or spelling rules, or Scripture verses, etc.), there is an underlying issue.** Always address from the outside in when trying to narrow down what is going on. Are they struggling with a physical issue that is either impeding them from taking in the information or processing for long term storage of memory? If everything is okay on the physical level, go to the character and spiritual level. Is this a pattern of laziness that is evident in other areas of their life? Is this a point of rebellion, disobedience, or stubbornness? Is it just plain childish foolishness with a disconnect between actions and consequences? If your child is dragging their feet in memorizing, please don't panic.

Note: We as parents tend to approach these situations with the mindset that our kids are "broken" or "bad" in some way, and we need to "fix" or "correct" them. We take it personally that we can't control the situation, which in turn feeds our insecurity. This type of thinking puts us in a "fix or punish" mindset, which puts everyone involved in a bad mood. No one likes to be approached in this way, including our children. It is much more beneficial to everyone involved if we approach it from the understanding that we are all developing and growing. Our children are not just learning the discipline to memorize, they are learning to be the boss of their will. We as parents can and should empathize with this, because we too, still battle to be the boss of our own wills. **Be aware, however, empathizing with someone is not the same thing as allowing them to take the easy path.** Memorization strengthens the brain. Not only are the facts beneficial, but the act of memorizing is also extremely brain-strengthening and benefits the child in every area.

Tips for Memory Work:

- **Don't overwhelm your student** with the number of facts to memorize.

- **Make flashcards and use them regularly.** I do periodically remind you to instruct your child to use their flashcards, but you decide if you want them to practice more often.

- **Play games.** For younger children, a game of matching works great.

- **Make memory work fun.** Lighten up. Celebrate victories.

- **Set an example.** Show your children that you are working on memory work as well.

- **Don't make math facts the only material you use for memory work.** Choose something from every area of learning to memorize — even if it is one sentence per week. Have recitation bees.*

- **Keep track of the facts** that your student knows by making a list each month or so of the ones they still need to work on memorizing.

Math Lessons for a Living Education is designed to not overwhelm the child or the parent with a ridiculous amount of memorizing. I have provided a chart for each of the levels, showing the important flashcards. Remember, this is not a race. Memorizing is not a one-shot deal; it takes time to become permanent. Allow your child to process at their own speed and facilitate the process.

You will notice that throughout the levels 3 and up, multiplication and division facts are reviewed and practiced in a variety of ways. Your student will work with them in their lessons, using vertical and horizontal problems, multiplication grids, copywork, and flashcards.

* I explain recitation bees in Part 4, Section 3.

Charts of Facts to Learn by Level

 Math Level 1

Day 71 (Lesson 15, Exercise 1)	2 + 2 = 4, 3 + 3 = 6, 3 + 1 = 4, 4 + 1 = 5
Day 76 (Lesson 16, Exercise 1)	4 + 4 = 8, 5 + 5 = 10
Day 88 (Lesson 18, Exercise 3)	[10's Family] 1 + 9 = 10, 2 + 8 = 10, 3 + 7 = 10, 4 + 6 = 10, 5 + 5 = 10
Day 116 (Lesson 24, Exercise 1)	3 + 4 = 7, 4 + 5 = 9
Day 123 (Lesson 25, Exercise 3)	2 + 3 = 5, 3 + 5 = 8
Day 126 (Lesson 26, Exercise 1)	Time Concepts Flashcards: refer to course schedule in Student Book for page number of Lesson 26, Exercise 1

Day 6 (Lesson 2, Exercise 1)	$4 + 1 = 5$, $1 + 4 = 5$, $2 + 3 = 5$, $3 + 2 = 5$
Day 10 (Lesson 2, Exercise 5)	24 hours = 1 day, 12 months = 1 year, 7 days = 1 week, Optional: 60 seconds = 1 minute 7 days of the week flashcards 12 months of the year flashcards
Day 13-14 (Lesson 3, Exercise 3-4)	Optional flashcard: 12 inches = 1 foot [from Lesson 12, Exercise 1] Subtraction facts of Math Level 2 (Make flashcards; split activity over two days)
Day 75 (Lesson 15, Exercise 5)	Doubles facts: refer to course schedule in Student Book for page number of Lesson 19, Exercise 5 of *Math Level 2*. I highly suggest using triangle flashcards for these.
Day 94 (Lesson 19, Exercise 4)	Money concepts flashcards: refer to course schedule in Student Book for page number of Lesson 19, Exercise 4 of *Math Level 2*.
Day 111 (Lesson 23, Exercise 1)	Measurement/weight concepts flashcards: 16 ounces (oz) = 1 pound (lb), Example of an ounce: a paper clip Example of a pound: a loaf of bread
Day 118 (Lesson 24, Exercise 3)	More measurement/weight concepts flashcards: 2 cups = 1 pint, 4 cups = 1 quart 2 pints = 1 quart, 8 pints = 1 gallon 4 quarts = 1 gallon, 16 cups = 1 gallon
Day 143 (Lesson 29, Exercise 3)	365 days = 1 year 52 weeks = 1 year

Move your finger from one number
to the next as you say it.

Move your finger from one number
to the next as you say it.

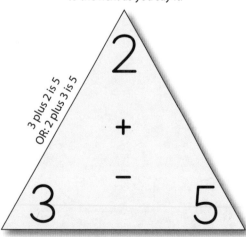

Like This: 5 take away 3 is 2.
OR: 5 take away 2 is 3.

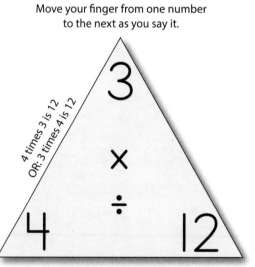

Like This: 12 divided by 3 is 4
OR: 12 divided by 4 is 3

Part 1, Addition/Subtraction Facts

Important note: If you are beginning the series at level 3, please take a few minutes to assess where your student is in the process of memorizing their addition and subtraction facts. There are two pages of addition and subtraction facts in Lesson 2 of *Math Level 3*. You may use these to see what facts your student has memorized. Simply watch them to see if they have to count to find the answer. Have them place a star next to the ones they need to memorize. Please take the time to create flashcards for the facts they need to memorize, and work on them often. Look through the chart showing the flashcards for *Math Level 2*. Notice especially the measurement facts learned in *Math Level 2*. If your child does not know these facts, please have them create flashcards for them, and work on them often until they are memorized.

In *Math Level 3*, the memorization of facts is a little different than in the other levels. Before multiplication and division are introduced in Lesson 12, the focus is to make sure the student has a grasp on the relationship between addition and subtraction. To cement this relationship in their minds, I have them learn the addition/subtraction fact families. Below, I have created a chart for memorization of the main fact families to be completed between Lesson 4 and 10.

Please Note: this is an *optional* assignment. If your student knows these well, please do not have them complete this.

Directions for making right-brain flashcards for addition/subtraction fact families: Study the example of the triangle flashcards on the previous page. Make one card for each family.

During Lesson 4	Family #1: 2 + 3 = 5, 3 + 2 = 5, 5 - 2 = 3, 5 - 3 = 2
During Lesson 5	Family #2: 3 + 4 = 7, 4 + 3 = 7, 7 - 3 = 4, 7 - 4 = 3
During Lesson 6	Family #3: 4 + 5 = 9, 5 + 4 = 9, 9 - 4 = 5, 9 - 5 = 4
During Lesson 7	Family #4: 2 + 8 = 10, 8 + 2 =10, 10 - 2 = 8, 10 - 8 = 2
During Lesson 8	Family #5: 3 + 5 = 8, 5 + 3 = 8, 8 - 3 = 5, 8 - 5 = 3
During Lesson 9	Family #6: 6 + 4 = 10, 4 + 6 = 10, 10 - 4 = 6, 10 - 6 = 4
During Lesson 10	Family #7: 4 + 7 = 11, 7 + 4 = 11, 11 - 4 = 7, 11 - 7 = 4

Math Level 3 Continued (Multiplication & Division Facts*)
[Optional schedule: Triangle right-brain flashcards]
Each row of facts on this page and continued on the next page represent one flash card.

During Lesson 13 Create one card per day	2 x 2 = 4, 4 ÷ 2 = 2 2 x 3 = 6, 3 x 2 = 6, 6 ÷ 2 = 3, 6 ÷ 3 = 2 2 x 4 = 8, 4 x 2 = 8, 8 ÷ 2 = 4, 8 ÷ 4 = 2 2 x 5 = 10, 5 x 2 = 10, 10 ÷ 2 = 5, 10 ÷ 5 = 2 2 x 6 = 12, 6 x 2 = 12, 12 ÷ 2 = 6, 12 ÷ 6 = 2
During Lesson 14 Create one card per day	2 x 7 = 14, 7 x 2 = 14, 14 ÷ 2 = 7, 14 ÷ 7 = 2 2 x 8 = 16, 8 x 2 = 16, 16 ÷ 2 = 8, 16 ÷ 8 = 2 2 x 9 = 18, 9 x 2 = 18, 18 ÷ 2 = 9, 18 ÷ 9 = 2 2 x 10 = 20, 10 x 2 = 20, 20 ÷ 2 = 10, 20 ÷ 10 = 2
During Lesson 15 Create one card per day	5 x 3 = 15, 3 x 5 = 15, 15 ÷ 5 = 3, 15 ÷ 3 = 5 5 x 4 = 20, 4 x 5 = 20, 20 ÷ 4 = 5, 20 ÷ 5 = 4 5 x 5 = 25, 25 ÷ 5 = 5 5 x 6 = 30, 6 x 5 = 30, 30 ÷ 5 = 6, 30 ÷ 6 = 5 5 x 7 = 35, 7 x 5 = 35, 35 ÷ 5 = 7, 35 ÷ 7 = 5
During Lesson 16 Create one card per day	5 x 8 = 40, 8 x 5 = 40, 40 ÷ 5 = 8, 40 ÷ 8 = 5 5 x 9 = 45, 9 x 5 = 45, 45 ÷ 5 = 9, 45 ÷ 9 = 5 5 x 10 = 50, 10 x 5 = 50, 50 ÷ 5 = 10, 50 ÷ 10 = 5 3 x 3 = 9, 9 ÷ 3 = 3 3 x 4 = 12, 4 x 3 = 12, 12 ÷ 3 = 4, 12 ÷ 4 = 3
During Lesson 17 Create one card per day	3 x 6 = 18, 6 x 3 = 18, 18 ÷ 3 = 6, 18 ÷ 6 = 3 3 x 7 = 21, 7 x 3 = 21, 21 ÷ 3 = 7, 21 ÷ 7 = 3 3 x 8 = 24, 8 x 3 = 24, 24 ÷ 3 = 8, 24 ÷ 8 = 3 3 x 9 = 27, 9 x 3 = 27, 27 ÷ 3 = 9, 27 ÷ 9 = 3 3 x 10 = 30, 10 x 3 = 30, 30 ÷ 3 = 10, 30 ÷ 10 = 3
During Lesson 18 Create one card per day	4 x 4 = 16, 16 ÷ 4 = 4 4 x 6 = 24, 6 x 4 = 24, 24 ÷ 4 = 6, 24 ÷ 6 = 4 4 x 7 = 28, 7 x 4 = 28, 28 ÷ 4 = 7, 28 ÷ 7 = 4 4 x 8 = 32, 8 x 4 = 32, 32 ÷ 4 = 8, 32 ÷ 8 = 4 4 x 9 = 36, 9 x 4 = 36, 36 ÷ 4 = 9, 36 ÷ 9 = 4

Math Level 3 Continued (Multiplication & Division Facts*)
[Optional schedule: Triangle right-brain flashcards]
Each row of facts on this page and continued on the next page represent one flash card.

During Lesson 19 Create one card per day	$4 \times 10 = 40$, $10 \times 4 = 40$, $40 \div 4 = 10$, $40 \div 10 = 4$ $6 \times 6 = 36$, $36 \div 6 = 6$ $6 \times 7 = 42$, $7 \times 6 = 42$, $42 \div 6 = 7$, $42 \div 7 = 6$ $6 \times 8 = 48$, $8 \times 6 = 48$, $48 \div 6 = 8$, $48 \div 8 = 6$ $6 \times 9 = 54$, $9 \times 6 = 54$, $54 \div 6 = 9$, $54 \div 9 = 6$
During Lesson 20 Create one card per day	$6 \times 10 = 60$, $10 \times 6 = 60$, $60 \div 6 = 10$, $60 \div 10 = 6$ $7 \times 7 = 49$, $49 \div 7 = 7$ $7 \times 8 = 56$, $8 \times 7 = 56$, $56 \div 7 = 8$, $56 \div 8 = 7$ $7 \times 9 = 63$, $9 \times 7 = 63$, $63 \div 7 = 9$, $63 \div 9 = 7$ $7 \times 10 = 70$, $10 \times 7 = 70$, $70 \div 7 = 10$, $70 \div 10 = 7$
During Lesson 21 Create one card per day	$8 \times 8 = 64$, $64 \div 8 = 8$ $8 \times 9 = 72$, $9 \times 8 = 72$, $72 \div 8 = 9$, $72 \div 9 = 8$ $8 \times 10 = 80$, $10 \times 8 = 80$, $80 \div 8 = 10$, $80 \div 10 = 8$ $9 \times 9 = 81$, $81 \div 9 = 9$ $9 \times 10 = 90$, $10 \times 9 = 90$, $90 \div 9 = 10$, $90 \div 10 = 9$

* You will notice what seems to be missing facts in each family. This is because your child has already created a card for that fact as part of another fact family. For example: 3×5 is missing from the 3's grouping because it was already included in the 5's grouping.

Math
Level 4

Day 50 [Lesson 10, Exercise 5]	Measurement facts: 8 quarts = 1 peck, 4 pecks = 1 bushel From least to greatest: pint, quart, gallon, peck, bushel
Day 122 [Lesson 25, Exercise 2]	Shapes flashcards: refer to course schedule in Student Book for page number of Lesson 25, Exercise 2 of *Math Level 4*

Day 61 [Lesson 13, Exercise 1]	Thinking Tools card: divisible by 2?
Day 62 [Lesson 13, Exercise 2]	Thinking Tools cards: divisible by 5? divisible by 10?
Day 63 [Lesson 13, Exercise 3]	Thinking Tools cards: divisible by 3? divisible by 9?
Day 65 [Lesson 13, Exercise 5]	Thinking Tools card: divisible by 4?
Days 91-95 [Lesson 19, Exercise 1-5]	Thinking Tools cards: #1-#9: refer to course schedule in Student Book for page number of Lesson 19, Exercise 1-5
Day 100 [Lesson 20, Exercise 5]	New concept card
Day 105 [Lesson 21, Exercise 5]	New concept card
Day 130 [Lesson 26, Exercise 5]	New concept cards #1 and #2
Day 135 [Lesson 27, Exercise 5]	3 Thinking Tools cards

Section #5: Oral Narration in Math?

The Why

One of the most common questions I receive about the *Math Lessons for a Living Education* series is "Why are there no tests? Don't I need a test to see what my kids know?" Although I do believe that test-taking is a necessary skill, I have found it to be counterproductive when it comes to young children and math (and every other subject). Most of us, when we know we have a test approaching, subconsciously switch to a different brain mode. Instead of engaging with the concepts in a creative/critical fashion, we begin to worry about what is on the test. Most of us, if we grew up in a traditional classroom setting, have been trained to study for the test, dump our stored-up rote memory information, and move on. Case in point…

Throughout my high school years, I made straight A's on my report cards. I even made excellent grades in Spanish (which I took for four years). I don't speak Spanish. I look at the few papers I have from those years, and the content is a foreign language. Although at the time, I could conjugate verbs in Spanish, write entire conversations in Spanish, and diagram sentences in Spanish, I was never required to actually speak it. I used a textbook that had me fill in the blanks and take tests. How many hours did I waste "doing Spanish" those four years? How many hours did I spend doing something that had absolutely no long-term value to me? And for what? So, I could sport an A+ on my report cards? I use my experience with Spanish as an example, but basically every other subject I took those years in a traditional classroom setting were the same. Read a textbook. Answer the end-of-the-chapter questions. Take a test. Forget it all within three months to a year. I learned to become good at taking tests, not at actually learning.

Homeschooling gives us the opportunity to use better methods of seeing what our children know.

This is where oral (and later, written) narration comes in. I have already written in-depth about the importance of teaching our children to train their brains' hemispheres to work together, and I've explained the importance of them being able to articulate their thoughts. (If you have not read about this, please go back to Part 1, Section 2 now.) This is what oral narration does for them. Oral narration requires many levels of thinking — far more than taking a test does. Remember, communication is a learned skill, and we, the parents, are meant to be gentle guides in the acquisition of it.

To orally narrate what they know about a particular concept, they have to mentally review what they have learned, choose what is important information, and place that information into the proper sequence. Next, they have to create sentences explaining the information in the same order that they have it arranged in their minds. Most of the time, I instruct the student to either use manipulatives or write on a whiteboard to show what they are saying. Oral narration is hard. In fact, on a scale of 1 to 10 — with 10 being extremely difficult — for most kids (and their parents), it's at least a 9. But it is worth doing. It's not just about the math concepts they are learning; it's about growing in mental acuity and agility. The more the student does it, the better they get at it because their brains are being trained and are more under control.

The How

If you are starting with a young child, simply have them tell back stories you have read to them, stories they have read to themselves, math concepts that you have worked through together, and anything else that your day brings along. Did you read the Bible together as a family? Informally have your child tell you about it. Watch a fun movie together on movie night? Ask your child about it a day or so later. Don't be all worried about them getting all of the details and plot twists — just let them retell it the way they remember it. If there is a hugely glaring discrepancy in their understanding, just matter of factly correct them by saying something like, "Actually, honey, this is what happened." Don't make it a big deal. Everyone misunderstands or misinterprets sometimes. If your child tends to be a perfectionist and is overly sensitive to correction or making a mistake, this will help them to realize that it is a normal part of life.

On a side note, I had a child who was extremely sensitive to making mistakes and didn't like to be corrected or criticized for anything. She would cry or sulk whenever she made a mistake, broke something, or even misspelled a word. I decided that I was going to come alongside her to help her with this particular weakness of her character instead of walking on eggshells around her in order not to hurt her feelings. I began by pointing out my own faults and mistakes. If I was writing on the whiteboard and spelled a word wrong, I said something like, "Oh my goodness, I completely misspelled that! Hmmm…I don't know how to spell it. Honey, would you look that word up for me in your dictionary and tell me how to spell it?" Or if I accidentally spilled something, I would apologize out loud to everyone around me and say, "Oh, I'm sorry! Did I get that on you? Please let me wipe it up before you step in it!" The more I took everyday situations and turned them into "failing forward" opportunities, the more all of my kids loosened up, and the perfectionistic one began emulating what I was modeling for her.

If you are starting oral narration with an older student (Jr. high and up), this transition into actually thinking through concepts and articulating them can be a little rough — I'm not going to kid you. My eldest child was a middle-schooler when I began using oral narration with my children. We were coming from a curriculum that tested everything the kids studied. It was a challenging transition for both of my older children. We began very small and built up, like this: if we were reading, I stopped every paragraph and had them take turns orally narrating what I had just read. When we first started with this, I had to reread the same paragraph multiple times before they could orally narrate (I thought I was doing to lose my mind, honestly). But we kept going, and it paid off.

After a month or so of stopping after each paragraph, I started reading a few paragraphs at a time before stopping. My children had learned that they had better pay attention because they didn't know who I was going to call upon. I also had them give me oral narrations about the books they were using for readers. We carried this method across our curriculum, with oral narrations for every subject, including math. My two older children were beginning to get into the bigger place value concepts. The use of oral narration exposed some glaring holes in their foundational understanding, so we backed up all the way to the beginning and began with the ones' place. Whenever I taught them a lesson or they read the lesson themselves, they knew that they better listen and read for understanding because, chances were, they were going to have to narrate and show what they had learned. **Oral narration nips lazy reading and listening in the bud and sets the student on the track for advanced critical reading skills needed in higher education.**

Although I have been asked numerous times for narration prompts to use, the value of oral narration comes from the parent's knowledge of what their own unique child has been learning, as well as their strengths, weaknesses, and learning style. Although I love children and want to help parents in their homeschooling journey, there is only so much I can do because I don't know your child. Oral narration requires the parent to plug into their child's learning journey and process. There is really no right or wrong way of asking for an oral narration. Simply ask the child to retell what they read, saw, heard, etc. The good news is, the more plugged in you become, the easier it is to assess where your child is in their journey.

Part 3

Teaching Instructions and Tips for Review Lessons — for those not starting at the beginning of the series

Section #1: Teaching Instructions and Tips by Level

Math Level K

Because families starting with *Math Level K* are starting at the very beginning of this series, they will be learning as they go along about the method and thought behind it. My only additional suggestion for these parents is to read this *Companion* in its entirety.

Math Level 1

Children at this age are just beginning to build numeric sense and spatial reasoning. To help them in this endeavor, I encourage parents to communicate with them about it. Point out examples of what they are learning in the world around them. (*Math Level 1* is full of a variety of nature connections with an emphasis on life cycles.)

Play games, enjoy hands-on activities (I've included a list of my favorites in Part 4, Section 3 of this *Companion*), and talk, talk, and talk!

Math Level 2

If you are beginning the *Math Lessons for a Living Education* series at this point without completing either of the previous volumes, read your student the story synopsis for *Math Level K* and *Math Level 1*, in Part 4, Section 5 of this *Companion*. Please read the entire *Teaching Companion* to equip yourself to teach this curriculum.

The first four Lessons in *Math Level 2* are a review of what was learned in *Math Level 1* and therefore, do not contain actual concept introduction. Here are some teaching tips for each of those lessons.

1. Place Value Village, Telling Time, Shapes, and Patterns:

- In preparation for constructing and teaching place value with Place Value Village, please read the section, in Part 2, Section 2 of this *Companion*, about teaching place value. Also, read the teacher's note and instructions in Lesson 1, Exercise 1 of *Math Level 2*, and watch the instructional video mentioned on that page. Explain to your child that you both will be embarking on a new learning journey. You may even want to read them portions of the section about place value in Part 2 of this Companion.

- In preparation for beginning copywork in math, determine if your child will do all or some of the copywork in these lessons. If your child is young or struggles with the mechanics of writing, I suggest that you copy the number in a highlighter and allow them to trace over it. This is a technique I used with several of my own children when they were younger until they built up the stamina needed to write for more extended periods.

- In preparation for telling time, make sure your child understands these time concepts: the hour and minute hands and their functions, how we tell time to the hour (the position of the minute hand pointing at 12 and the hour hand at whatever number the hour is). Your student will be constructing a clock in the next lesson.

- In preparation for the shapes and patterns, talk through the shapes and patterns on Lesson 1, Exercise 3 of *Math Level 2*. Make sure your child knows the name of each of the shapes — square, circle, triangle, and rectangle. Make sure they can tell you what makes up each shape. For example: a square has four equal straight sides.

2. **Addition (horizontal and vertical), shapes**

- **Flashcards** – please read the entire section in this Companion about right-brain flashcards.

- Make sure your child understands the concept of addition. Use small counting items to work with the concept for a few minutes before moving into the written work. Practice the addition rule: you can place the addends of an addition equation in any order and still get the same answer (sum).

- Practice the days of the week together. Work together to create the calendar using the instructions on Exercise 2, Day 7. Practice the months of the year.

3. Subtraction

- Use small counting items to work with the concept of subtraction before doing the written work in these lessons. Practice the subtraction rule: unlike addition, in subtraction, it matters which number is first. The larger number comes first and the smaller number is taken away from it. For example: $10 - 7 = 3$ not $7 - 10$. Have your student practice and explain this concept to you with small counting items.

4. Writing numbers to 100 and simple fractions

- Work one-on-one with your student to review all numbers to 100. Take time to play some games with the 100's chart in the chart section of *Math Level 2*.

- Spend some time reviewing what makes a true fraction (all equal parts that make up a whole). Draw some shapes and allow your student to practice dividing them into true fractions.

Math Level 3

If you are beginning the *Math Lessons for a Living Education* series at this point without completing either of the previous volumes, read your student the story synopsis for *Math Level K – Math Level 2*, in Part 4, Section 5 of this *Companion*. Please read the entire *Teaching Companion* to equip yourself to teach this curriculum.

Please look through the *Math Level 3* Chart of Facts to Learn on pages 27-29 in Part 2, Section 4 of this *Companion*.

The first six lessons in *Math Level 3* are (mostly) a review of what was learned in *Math Level 2* and therefore do not contain actual concept introduction. Here are some teaching tips for each of these lessons.

1. **Review of Place Value, Odds and Evens, Counting by 2s, 5s, and 10s**

- In preparation for constructing and teaching place value with Place Value Village, read the section, in Part 2, Section 2 of this *Companion*, about teaching place value.

- Watch the tutorial on Place Value Village. (Link in Part 2, Section 2 of this *Companion*, in the section about teaching place value.)

- Set aside a block of time (a couple of days at least) for your student to practice the concepts in this lesson.

- Even and Odd — use small counting items to review this concept. Say to your student: all even numbers end in 0, 2, 4, 6, and 8. Even numbers can be divided evenly into groups of 2 without any left over. Odd numbers end in 1, 3, 5, 7, and 9. Odd numbers cannot be divided evenly into groups of 2 without having one left over. Use the 100's chart from the Manipulatives section of *Math Level 3* to play the Odds & Evens Game from Part 4, Section 3 of this *Companion*.

- At the end of Lesson 1, your student will be creating a skip counting poster. This is NOT an optional assignment. Skip counting is a pre-multiplication activity and one that will strengthen your student's ability to memorize the multiplication facts more quickly when they get to that concept in coming lessons.

2. **Review of Money, Clocks, Perimeter, Addition/Subtraction Facts**

- If needed, spend some time with your student reviewing money and time concepts. Make sure they are comfortable with the value of each coin before assigning them the written work.

- If needed, spend some time with your student reviewing the concepts of telling time. Review how many hours are in a day, how many minutes are in an hour, and the minute hand and hour hand with their functions.

- If your student is not familiar with perimeter, follow these directions: use a large area, such as a room or a large area outside, and have them walk off the outer edge. Explain to them that they just walked the perimeter of that room/space. To find the perimeter of a shape, simply measure each side and add all of them together. In Lesson 2, Exercise 3 of *Math Level 3*, there are examples of a triangle and a square. Work with your student to find these perimeters by adding all of the sides together.

- If needed, review and practice the addition and subtraction facts in Lesson 2, Exercise 4 and Lesson 2, Exercise 5 before having your student complete those pages. Take note about which facts they need to work on memorizing.

3. **Review of Addition, Including Carrying and Tally Marks**

- Addition with carrying — If needed, review this concept with your child. Use the problems in Lesson 3, Exercise 1 and work through these steps.

➤ On a piece of paper or a whiteboard, draw what is shown below.

➤ Write the problems as shown. Explain that the numbers in the one's house represent groups of one, while the numbers in the tens' house represent groups of ten. Add the numbers in the ones' house: 5 (it's not more than 9, so there is no need to carry). Add the numbers in the tens' house: 9. So the answer is 95, which is 9 groups of 10 and 5 groups of 1.

➤ Next, do this problem. Repeat that the numbers in the one's house represent groups of one, while the numbers in the tens' house represent groups of ten. Add the ones place first: 2 + 9 = 11. This means that there are too many ones to stay in the ones' house. The number 11 is 1 group of ten and 1 group of one. We need to carry the 1 group of ten up to the tens' house to be added with the rest of the groups of ten. The final answer is 151.

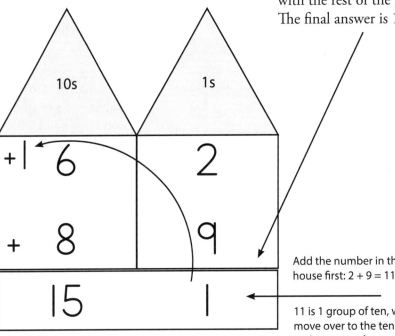

Add the number in the ten's house, including the group of ten carried from the ones' house, next: 1 + 6 + 8 = 15

The answer to the whole problem: 62 + 89 = 151

Add the number in the ones' house first: 2 + 9 = 11

11 is 1 group of ten, which we move over to the tens' house, and 1 group of one, which we leave in the ones' house.

➤ If needed, practice **tally marks** with your student. Explain to them that when writing tally marks for five or higher, the first four tally marks are placed side by side with the fifth drawn diagonally across them (look at the first one in Lesson 3, Exercise 2 of *Math Level 3*). Practice doing tally marks for things around the house.

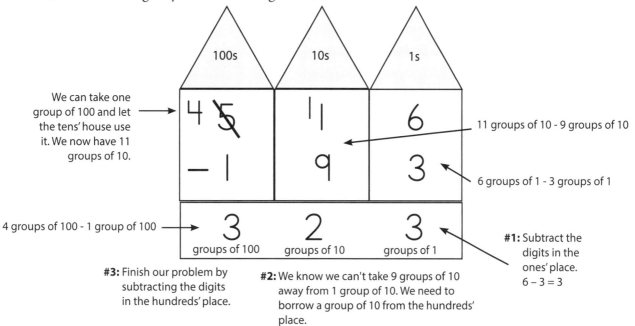

We can take one group of 100 and let the tens' house use it. We now have 11 groups of 10.

11 groups of 10 - 9 groups of 10

6 groups of 1 - 3 groups of 1

4 groups of 100 - 1 group of 100

#1: Subtract the digits in the ones' place. 6 − 3 = 3

#3: Finish our problem by subtracting the digits in the hundreds' place.

#2: We know we can't take 9 groups of 10 away from 1 group of 10. We need to borrow a group of 10 from the hundreds' place.

4. Review of Subtraction, Including Borrowing Concepts

- If needed, review the process of subtraction with carrying following the steps shown above.

5. Review of Measurement, Fractions, Thermometers, and Graphs

If needed, review how a thermometer is designed. Show your student how each of the little lines between the numbers stands for two degrees, so we count by twos when we are reading the temperature.

For the graph in Lesson 5, Exercise 1 of *Math Level 3*, your student will be filling in the temperatures for five days. Simply have them write a range of temperatures that coincides with the temperature you have experienced in those five days. (For example: if the temperatures are between 62 and 68 degrees as a high, have them write 62, 64, 66, 68, and 70 on the lines.) Each day your student will draw a dot on the vertical line for that day. Make sure they understand that they are drawing the dot at the temperature for that day.

Explore the bar graph together.

Make sure your student is well on their way to learning the measurement facts on Lesson 5, Exercise 2.

6. Solving Word Problems.

Students who have been using this math series from the beginning will be familiar with the steps of solving word problems. However, if you are entering the series at *Math Level 3*, please discuss the steps outlined at the top of Lesson 6, Exercise 1 and work through the problems on the bottom of the page. Explain to your child that solving word problems is like solving a puzzle. You have to organize the information and ask yourself questions before solving. Also explain that this is a concept and skill that requires time, patience, and a lot of practice.

If you are beginning the *Math Lessons for a Living Education* series at this point without completing any of the previous volumes, read your student the story synopsis for *Math Level K – Math Level 3*, in Part 4, Section 5 of this *Companion*. **Please read the entire *Teaching Companion* to equip yourself to teach this curriculum.**

- Please look through the *Math Level 4* Chart of Facts to Learn on page 29 in Part 2, Section 4 of this *Companion*.

- The first six Lessons in *Math Level 4* are (mostly) a review of what was learned in *Math Level 3* and therefore do not contain actual concept introduction. Here are some teaching tips for each of those lessons.

1. Addition and Subtraction Concepts

If your child needs to review and practice the processes of carrying and borrowing, please look at the examples shown in the previous section for *Math Level 3*. Please make sure your child understands fully. Allow them time to practice these until they are comfortable.

For the missing number assignment in Lesson 1, Exercise 2 of *Math Level 4*, if needed, instruct your student to show you the first column of equations using manipulatives and explain what they are doing.

If needed, take some time to walk through the concept of telling time. Allow your student time to practice showing how to tell time.

If needed, take time to explain and practice using thermometers. Explain to your child that each of the lines between the numbers stands for two degrees. Therefore, we count by 2s when reading a thermometer.

Before solving the word problems in Lesson 1, Exercise 4, discuss these steps:

- Read the problem carefully.
- Ask "what is the question?"
- Circle the numbers you will need to use to solve the problem.
- Think it through. What operation will you need to use to find your answer?
- Solve. Think through your answer to see if it makes sense.

2. Review of Place Value, Estimation and Rounding

If needed, use Lesson 2, Exercise 1 and Lesson 2, Exercise 2 of *Math Level 4* to teach and/or review the concepts of rounding and estimation. Here's how: on a whiteboard or a piece of paper, draw a simple number line with numbers 0 - 100 counting by 10s. Leave a space between and make a small mark halfway between each number (study the illustration below). In Lesson 2, Exercise 1, start at the top of the page and say, "27 is between what two 10s?" (20 and 30). Look at the last digit in each number. Is it 5 or more? Then it is closer to the bigger 10. (27 is closer to 30 because 7 is more than 5.) Work through all of the assignments on this page like this. (Study how to create a number line for each section below.)

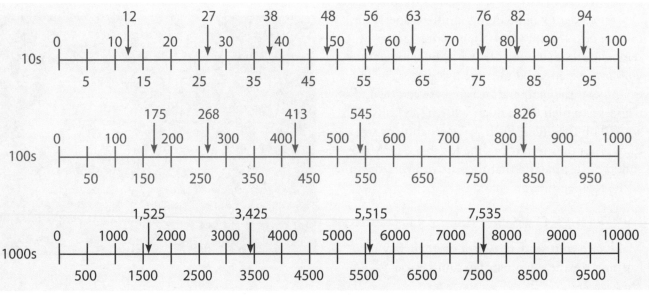

- Rounding to the tens' place: look at the number in the ones' place; less than 5, round down to the smaller ten, 5 or more, round up to the larger ten.

- Rounding to the hundreds' place: look at the number in the tens' place; less than 5, round down to the smaller hundred, 5 or more, round up to the larger hundred.

- Rounding to the thousands' place: look at the number in the hundreds' place; less than 5, round down to the smaller thousand, 5 or more, round up to the larger thousand.

- To round and estimate in addition (Lesson 2, Exercise 2 of *Math Level 4*), simply round each number and add together. The answer is the estimated sum. If needed, allow your child to use a number line like the one in the previous activity.

If needed, inform your child that a comma is placed in a number every three digits from the right.

If needed, review place value through the millions' place. Work through each of these numbers with your student:

- 5,700,104 5 groups of 1,000,000, 7 groups of 100,000, 1 group of 100, and 4 groups of 1

- 472,186 4 groups of 100,000, 7 groups of 10,000, 2 groups of 1,000, 1 group of 100, 8 groups of 10, and 6 groups of 1

If needed, review all multiplication concepts with your student. Make sure they understand by using manipulatives to show all of the facts in the grid in Lesson 3, Exercise 1 of *Math Level 4*. Talk through these concepts: 2 x 4 = 8. This is the same as saying two groups of 4 OR four groups of 2.

If needed, review all division concepts. Use Lesson 4, Exercise 1 to work through the relationship between multiplication and division. Use manipulatives to show them the multiplication facts on the left side of the page. Use the same manipulatives to show them the division facts. Allow them time to work through these two concepts until they can show you without coaching.

Measurement facts. If needed, work with your student to review the measurement facts using the chart in the Manipulatives Section of *Math Level 4*.

If your student has not done an assignment like those in Lesson 5, Exercise 4, work through it with them. This is a unique and helpful way of learning the connection between division and fractions.

Roman numerals. If your student does not have the basic Roman numerals memorized, use the ones in Lesson 6 to create a learning chart or flashcards to memorize them.

Perimeter. If needed, work through the perimeter problems in Lesson 6. Explain: the perimeter of a shape is found by adding all of the sides' lengths together.

Math Level 5

If you are beginning the *Math Lessons for a Living Education* series at this point without completing any of the previous volumes, read your student the story synopsis for *Math Level K – Math Level 4*, in Part 4, Section 5 of this *Companion*. Please read the entire *Teaching Companion* to equip yourself to teach this curriculum. There are a number of helpful charts in the back of *Math Level 5* that are used to help the student work through these review lessons and the lessons that follow them. For more information on these charts, please read the next section.

The first six lessons of *Math Level 5* are a review that covers solving addition, subtraction, word problems, multiplication and division concepts, geometry, measurement, and fraction concepts.

For examples of how to teach addition with carrying and subtraction with borrowing, go to the teaching tips for *Math Level 3*, earlier in this section. Make sure your student understands these concepts well.

For instructions on reviewing the concepts of multiplication and division, go to the teaching tips for *Math Level 4*, earlier in this section. Make sure your student understands these concepts well.

If your student needs guidance through the geometry assignment in Lesson 3, please make sure they are using the Geometry Chart from the Manipulatives Section of the book. They are assigned for your child to narrate to you what they are doing. Please make sure they understand these concepts.

Measurements. Use the measurement chart in the Manipulatives Section to work through these assignments.

For word problems, review these steps to solving:

- Read the problem carefully.
- Ask "What is the question?"
- Circle the numbers you will need to use to solve the problem.
- Think it through. What operation will you need to use to find your answer?
- Solve. Think through your answer to see if it makes sense.

Fractional concepts. If needed, have your student draw the problems as part of Lesson 5, and solve them together.

$$3\frac{1}{6} + 1\frac{3}{6} = 4\frac{4}{6}$$

If needed, to teach the type of problems in the bottom half of Lesson 5 Exercise 1, show your student this illustration.

$$\frac{1}{5} \text{ of } 10 = 2$$

Decimal concepts learned so far are all explained thoroughly and reviewed in Lesson 6. Please make sure your child understands these.

Math Level 6

Although *Math Level 6* begins with a review, it is not set up like *Math Levels 2-5*, where the first 4 to 6 weeks were review from the previous volumes. *Math Level 6* has fully integrated instructions that start the student at the very beginning of each concept and take them quickly up to the more difficult levels of that concept.

Math Level 1

In the Manipulatives Section of *Math Level 1*, you will find a comprehensive list of where in the book each chart is referenced and/or used.

The houses and counting mat for the Place Value Village are included. Please remove these pages from the book. Cut out the houses, (optional) color them, (optional) laminate them, and stick them on the side of a container. Another option is to simply lay them in order on the table in front of your child. For more information on using the Place Value Village, please read Part 2, Section 2 of this *Companion*. The most important thing I want you to remember is this: place value can be taught in a variety of ways; the Place Value Village is meant to be used as a tool. **Please read** the entire section in this *Companion* about teaching and learning place value.

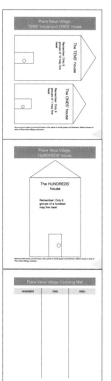

Also included in this section are addition mats. The goal for these is to add another way of practicing the concepts of addition. Please remove this page from the book, and either slip it into a page protector or cover it with some type of laminating plastic. The idea is to make it a wipeable surface. To use these mats: simply have your child use counting items and a wipeable marker to create addition equations.

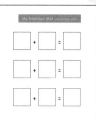

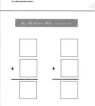

For your convenience, there is a page of number cards. To prepare this page, remove it from the book, cut the numbers apart and place them in an envelope. You may want to laminate them

for sturdiness. If you decide to laminate them, cut them out first and line them up with a space between them to make it easier to cut them apart again. Your child will be instructed to use these in multiple lessons throughout the book.

This is your child's 100's chart. This page needs to be removed from the book and laminated or slipped into a page protector. Your child will be using this chart throughout the course.

This book also includes the days of the week cards. Follow the directions you used for the number cards.

This is a clock template. Directions for assembly are in *Math Level 1*.

The manipulatives in this section also include an example of a right-brain flashcard made with an optional story.

Please Note: there is an answer key for *Math Levels 1* and *2* available for download at this link:

https://www.masterbooks.com/classroom-aids

In the Manipulatives Section of *Math Level 2*, you will find a comprehensive list of where in the book each chart is referenced and/or used.

This includes the houses and counting mat for the Place Value Village. Please remove these pages from the book. Cut out the houses, (optional) color them, (optional) laminate them, and stick them on the side of a container. Another option is to simply lay them in order on the table in front of your child. For more information on using the Place Value Village, please read Part 2, Section 2 of this *Companion*. The most important thing I want you to remember is this: place value can be taught in a variety of ways; the Place Value Village is meant to be used as a tool. **Please read** the entire section in this *Companion* about teaching and learning place value.

This section also contains a page of Hundreds Counters. When your child progresses to the Thousands' House, you will be instructed to use these counters instead of having them deal with that number of counting items. Simply remove the page from the book and cut out the counters. You may want to laminate them for sturdiness.

There is a 100's chart for your child. This page needs to be removed from the book and laminated or slipped into a page protector. Your child will be using this chart throughout the course.

A page of helpful number cards can also be found. To prepare this page, remove it from the book, cut the numbers apart and place them in an envelope. You may want to laminate them for sturdiness. If you decide to laminate them, cut them

out first and line them up with a space between them to make it easier to cut them apart again. Your child will be instructed to use these in multiple lessons throughout the book.

The Larger Addition Mat and the Larger Subtraction Mat, the Horizontal and Vertical Addition Mats for smaller problems (not pictured), are also included. Using these mats is simply another way of giving your child a way to practice these concepts. Remove them from the book, laminate them and use them to practice addition and subtraction concepts.

In *Math Lessons for a Living Education*, I try to give the student multiple ways of practicing their addition, subtraction, multiplication, and division facts. For this level, this includes the Addition Fact Sheet, Subtraction Fact Sheet, and the Doubles Families Fact Sheet are some of these ways. Simply remove the pages from the book, laminate and allow your student to practice these facts with a wipeable marker.

There is an example of a right-brain flashcard made with an optional story.

A reproducible calendar page is included, and you will need to make 12 copies. Your child will be creating their own calendar throughout the course.

Please Note: there is an answer key for Math Levels 1 and 2 available for download at this link:

https://www.masterbooks.com/classroom-aids

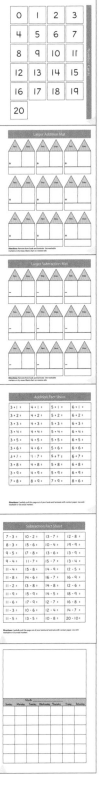

Math Level 3

In the manipulatives of *Math Level 3*, you will find a comprehensive list of where in the book each chart is referenced and/or used.

Included in the manipulatives are the houses and counting mat for the Place Value Village. Please remove these pages from the book. Cut out the houses, (optional) color them, (optional) laminate them, and stick them on the side of a container. Another option is to simply lay them in order on the table in front of your child. For more information on using the Place Value Village, please read Part 2, Section 2 of this *Companion*. The most important thing I want you to remember is this: place value can be taught in a variety of ways; the Place Value Village is meant to be used as a tool. Please read the entire section in this *Companion* about teaching and learning place value.

There is a page of Hundreds Counters. When your child progresses to the Thousands' House, you will be instructed to use these counters instead of having them deal with that number of counting items. Simply remove the page from the book and cut out the counters. You may want to laminate them for sturdiness.

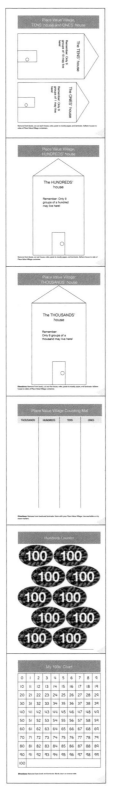

Also included is your child's 100's chart. This page needs to be removed from the book and laminated or slipped into a page protector. Your child will be using this chart throughout the course.

In *Math Lessons for a Living Education*, I try to give the student multiple ways of learning and practicing their multiplication and division facts. The Multiplication Facts and Division Facts for Copywork included are one of those ways. As your student works on memorizing the multiplication and division facts, simply have them choose several facts or fact families to use as copywork. It is amazing how we remember what we write.

There is also is a Multiplication Chart with facts from 0 through 12. Please laminate this chart for sturdiness and allow your student to use it when learning new patterns and concepts concerning multiplication and division.

In the Manipulatives Section, there is a small replica of the Place Value Village. Although students at this age have been at least exposed to place value, please make sure you read the entire section in this *Companion* about teaching it.

There are also instructions for the hands-on Fraction Kit project. Your child will be instructed when to create and use this kit while they are working through the course.

This level also presents a game for learning and practicing equivalent fractions.

A fraction game spinner along with the directions to put it together is also included.

There are "Break it Down Cards" showing the step-by-step solutions for several of the major concepts taught in *Math Level 4*.

Fraction/Decimal and Fraction/Decimal/Percent are also provided. These give the student a place for hands-on exploration and practice of these concepts. Please remove these pages from the book, laminate them, and allow your student to use them with a wipeable marker. The student will be instructed throughout the course when to use them, but they may also choose to use them during other times for more practice.

Another convenient part is the measurement charts. This page should be removed from the book, laminated and kept handy for use throughout the course.

Math Level 5

In the Manipulatives Section of *Math Level 5* are all of the charts that need to be removed from the book and either slipped into page protectors or laminated. The instructions on how and when to use these charts are in the lessons of the book.

This includes a special Long Division Practice Mat. This page should be removed, laminated, and kept in a safe place. Have your child use this once or twice a week for practice in long division.

Math Level 6

The Manipulatives Section of *Math Level 6* contains charts which will be used throughout the course. They should be removed from the back of the book and laminated.

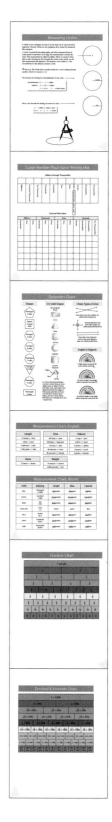

Part 4

Miscellaneous Resources

Section #1: Frequently Asked Questions

Question: How does *Math Lessons for a Living Education* scope and sequence and "speed of teaching" compare to other major, well-known traditional curriculums?

Answer: The overall scope and sequence (from K - 6) covers the same concepts covered in traditional math curriculum. Your student will cover everything they need to build a firm foundation of math concepts and will be ready to move into pre-algebra after *Math Level 6.* The "speed of teaching" is basically the same as most other curriculums. There may be slight variations from concept to concept, but not enough to say that it is faster or slower in an overall sense.

Question: Does *Math Lessons for a Living Education* use a spiral approach or a mastery approach?

Answer: This curriculum is actually a unique mixture of both. It is mastery in the sense that the student is required and expected to know the concepts. It is spiral because there is a consistent review given throughout the program to assist them in achieving the mastery needed.

Question: Is *Math Lessons for a Living Education* Common Core?

Answer: No.

Question: Why is *Math Lessons for a Living Education* called "math lessons with a flavor of Charlotte Mason"? Charlotte Mason did not use stories to teach math; she said that math was a language unto itself.

Answer: When I wrote this curriculum, I was very much aware of how Charlotte Mason taught math; I had tried it with my own children and had realized that I preferred to use many of the same techniques I had adapted from her method for other subjects in my teaching of math. With a little bit of adjustment here and there, I realized that the practice of oral narration, copywork, and living stories added a dimension to math that we had never had before. It lifted the flat symbols and equations off of the page and brought them to life in a joy-filled way. Why not bring the study of the living God for a living education into the study of Mathematics? Why not weave a delightful, character-building story into the concepts? Why not use the study of mathematics to disciple a child's heart for Christ, showing them the amazing display of His character revealed through His creation — numbers? I give credit where credit is due — Charlotte Mason is one who originally brought narration, copywork, and many other aspects that I have adapted to teaching math to the educational world. Thus, the "flavor of the Charlotte Mason method."

Question: Is *Math Lessons for a Living Education* enough by itself?

Answer: Short answer, yes.

Longer explanation: *Math Lessons for a Living Education* series was designed to be a tool to help teach our children to *think*, not just *do*. The world is full of people who *do*, but don't necessarily *think*. The one major difference between leaders and non-leaders is this: leaders know how to think. In fact, leaders schedule regular think-time. I believe with all of my heart that the children we are homeschooling are important leaders in God's future army. It was to that end that I created this curriculum. I didn't just imagine your tousle-haired kiddos gathered around your dining room table or sprawled across your couch while they worked away on their math exercise. I pictured them in the future. I asked myself, "How can I add value to these kids' mom's life? How can I help her facilitate their education and help them become everything God wants them to be? How can I come alongside her and help her teach them to *think*, not just *do*?"

Math Lessons for a Living Education is meant to not burden the student or the parent with excessively long lessons full of busywork. Where traditional math curriculums use pages of worksheets to drill the concepts into the child's brain, *MLFLE* uses a wide variety of activities. Yes, there are worksheets, but there are also hands-on activities that add movement to the learning process. Among other activities, your child will sew simple projects (no machine required), measure and cook simple but tasty recipes, create bird feeders out of stale bagels, and study and chart the birds that visit them. They will be invited to create their very own Mexican Fiesta (in Level 5) with instructions on how to plan, shop (if necessary), decorate, and invite the family to the finished shindig!

Many children have gotten so attached to the characters of the story that they have written me letters begging me to continue the series so they can hear "the rest of the story." I've received thank you cards from families who once "hated math so much" but now love it better than anything. At the very top of the list of victories are the eternal ones: children have given their hearts and lives to Christ as they have connected with the Biblical worldview aspect woven through the stories and concepts in this series. If you are looking for a math workbook to give your student having the goal of seat-work and drill, with little to no interaction with you as the parent, then I suggest that you continue your math curriculum shopping — this one will not fit your needs. If you are looking for a curriculum that will encourage your student to connect with you and with their learning process, you are holding it in your hands. Welcome.

From Math Level 1

In Lesson 5 of *Math Level 1*, Grandma makes the children her delicious blueberry pancakes. The recipe is not included in the book, so I thought I'd share it with you here. Whenever I give one of "Grandma's recipes," it is a recipe that I have created (or adapted extensively) and used over the years of cooking for my own family. This particular recipe can be made gluten-free by substituting a cup-to-cup gluten-free baking blend (our favorite for pancakes and waffles is the Pillsbury Best Multi-Purpose Gluten-Free Flour Blend™).

Grandma's Yummy Blueberry Pancakes
(Makes 10 pancakes — this recipe doubles well)

Ingredients:

- ☐ 2 eggs
- ☐ 1 cup of flour
- ☐ ¾ cup of milk (can substitute unsweetened, plain almond milk if desired)
- ☐ 1 Tbsp. sugar
- ☐ 2 Tbsp. vegetable oil (can substitute 1.5 Tbsp. melted butter if desired)
- ☐ 2 tsp. baking powder (be sure to sift out any clumps — you don't want any bitter lumps in your batter)
- ☐ ½ tsp. salt.
- ☐ 1 cup of blueberries — wild ones are the best (If using frozen berries, place them in a colander and run warm water over them until they are thawed. Drain before adding to the batter.)

Directions:

1. In large bowl, whisk together eggs, milk, and oil.
2. In a separate bowl, combine flour, sugar, baking powder, and salt.
3. Slowly add wet ingredients to dry, stirring all the while.
4. Stir only until mixed well. Do not over-stir.
5. Carefully fold in blueberries.
6. Pour into 3-4 inch rounds on a hot, greased griddle or frying pan over medium/high heat and flip when little bubbles begin to form (the other side will be golden).
7. Enjoy with butter and maple syrup.

In Lesson 24 and Lesson 25, Grandma is making biscuits and stew. The recipes are not included in the book, so I thought I'd share them with you here. My children loved helping make biscuits.

Grandma's Buttermilk Biscuits (10-12 biscuits)

Ingredients:

- ☐ 2 cups all-purpose flour (can substitute with 1¾ cup Pillsbury Best Multi-Purpose Gluten-Free Flour Blend™)
- ☐ 1 Tbsp. baking powder (well sifted to remove lumps)
- ☐ ¼ tsp. baking soda (well sifted to remove lumps)
- ☐ 1 Tbsp. sugar
- ☐ ½ tsp. salt
- ☐ 6 Tbsp. butter (cold and cut into chunks)
- ☐ ¾ cup buttermilk*

*If you do not have buttermilk on hand, you can make it! Pour ¾ cup cold regular milk into a cup measure, and add about 1.5 tsp. lemon juice. Stir and let sit for a few minutes before adding ¾ cup of it to your recipe.

Directions:

1. Preheat the oven to 425°. Combine all dry ingredients in a bowl.
2. Place cut-up butter on top of the dry ingredients and use a pastry cutter to cut it into the dry mixture. Do this until the butter is cut up into small pieces and is mixed throughout dry ingredients.
3. Pour small amounts of buttermilk into the mixture, gently kneading it with your hands to mix it.
4. Sprinkle a small amount of flour on a clean countertop and transfer the dough out onto it.
5. Gently work it with your hands — do not over-handle or you will end up with hard, leathery biscuits.
6. Gently pat the dough into a rectangle that is about ½ inch thick.
7. Dip a biscuit cutter or the edge of a drinking glass (about 2.5-3 inches in diameter) in flour and use it to cut out your biscuits. Cut them out as close to each other as possible. Gather up remaining dough, pat it down into ½ inch thickness again and cut out as many biscuits as possible.
8. Place biscuits touching each other on an ungreased baking sheet.
9. Bake at 425° for about 15 minutes (until they are golden on top and flaky in the middle).
10. Remove from oven, let cool on sheet for a few minutes then transfer each biscuit to a cooling rack. Brush with a little melted butter if desired.
11. Optional: Cover with a clean dish towel to keep the flies off! Serve with stew — recipe on the next page.

Grandma's Hearty Beef Stew
(Serves 6)

Ingredients:

- ☐ 1 to 1.5 lbs. of stew meat (may substitute ground beef, venison, or any other red meat)
- ☐ 2 cups carrots peeled and sliced into thick medallions
- ☐ 2 cups potatoes (red or yellow are best) scrubbed and cubed
- ☐ 1 small yellow or white onion, chopped fine
- ☐ 1 clove of garlic, minced
- ☐ ½ small can of tomato paste
- ☐ 1 tsp. garlic salt
- ☐ 1 tsp. thyme
- ☐ 1 bay leaf (remove before serving)
- ☐ Salt and pepper to taste
- ☐ ¼ cup beef bouillon and 1.5-2 quarts of water (warmed and mixed to make beef broth) or 1.5-2 quarts of beef broth
- ☐ 1.5-2 Tbsp. cornstarch for the roux

Directions:

1. Prepare carrots, potatoes, onion, garlic, and beef broth (if using beef bouillon). Set aside.
2. In a large skillet, brown stew meat (or alternative meat) until it is cooked rare and sprinkle with garlic salt while cooking.
3. Transfer meat and drippings into a large kettle.
4. Add the broth, thyme, onion, garlic, and tomato paste.
5. Cook over medium heat for 10-15 minutes or until boiling.
6. Add carrots and potatoes and continue to cook until they are beginning to soften.
7. Add bay leaf.
8. Add salt and pepper to taste.
9. Make the roux: Dip out 1.5 cups of the hot broth (try not to get anything but broth) and place in a small glass mixing bowl.
10. Stir in cornstarch with a fork, adding small amounts of it to the broth at a time. The mixture will begin to thicken.
11. Slowly add roux to the stew, stirring continually.
12. Continue to cook the stew over medium-low heat (stirring occasionally) for another 10-15 minutes or until the potatoes and carrots are tender. Stew may continue to thicken until served.
13. Remove the bay leaf before serving. Serve with Grandma's Buttermilk Biscuits.

This recipe in found in Lesson 26 of *Math Level 1*.

Grandma's Scruptdelicious Oatmeal
(Makes 6-8 servings)

Ingredients:

- ☐ 4½ cups of water
- ☐ 1 tsp. salt
- ☐ 1 tsp. pure vanilla extract
- ☐ ½ Tbsp. cinnamon
- ☐ ¼ tsp. nutmeg
- ☐ 3 cups whole rolled oats
- ☐ (optional, but so yummy!) ¼ cup raisins

Directions:

1. Stir everything together in a medium-sized pot and bring to boil over a medium heat.
2. Once at a boil, turn down to low, stirring occasionally, and let simmer until oats are tender.
3. Serve with sliced banana and a sprinkling of light brown sugar, or, if you prefer, honey.
4. Yummm! Warm and sweet!

From Math Level 2

Feed the birds! *Math Level 2* includes two recipes for creating edible bird feeders.

Wildlife Energy Muffins

You will need:

- ☐ 1 cup chunky peanut butter
- ☐ 1 cup pure rendered suet or vegetable shortening
- ☐ 2½ cups coarse yellow cornmeal
- ☐ Seeds, raisins, or other dried fruit and roasted peanuts
- ☐ Pipe cleaners

Directions:

1. Mix peanut butter, suet, and cornmeal together.
2. Stir in seeds, fruit, and nuts.
3. Make "muffins" by placing the mixture into a muffin tin.
4. Sprinkle seeds on top.
5. Fold a pipe cleaner in half and push both ends into each muffin to act as a hanger.
6. Place the tin in the freezer to harden.
7. Once hardened, hang the muffins from a tree.

Bagels for the Birds

You will need:

- ☐ 1 bag of bagels (old, stale — but not moldy — work best)
- ☐ 1 jar of plain peanut butter
- ☐ 1 bag of birdseed
- ☐ 1 roll of ribbon (cloth or gift-wrapping ribbon)

Directions:

1. Split bagels lengthwise, and let them harden overnight.
2. Tie lengths of ribbon through each bagel hole.
3. Spread peanut butter over both sides of each bagel slice.
4. Sprinkle with birdseed.
5. Hang bagels throughout your backyard.

In Lesson 14 of *Math Level 2*, the family is enjoying a breakfast which includes baked oatmeal with strawberries. The recipe is not included in the book, but I thought I'd include it here. This is a protein-packed alternative for regular oatmeal cooked on the top of the stove.

Baked Oatmeal with Strawberries (or other fruit) — Serves 5 or 6

Ingredients:

- ☐ 3 large eggs or 4 smaller ones (may substitute 2.5 cups of egg whites for a fat-free version)
- ☐ 1 cup milk (may substitute 1 cup of unsweetened almond milk for dairy-free version — leave out the salt in the recipe)
- ☐ ½ cup honey (raw is best) at room temperature (may substitute with ½ cup of unsweetened apple sauce or appropriate amount of preferred sugar replacement)
- ☐ 4 cups old-fashioned rolled oats
- ☐ 1 Tbsp. cinnamon
- ☐ 1 tsp. vanilla extract
- ☐ ½ tsp. nutmeg
- ☐ 1½ tsp. salt
- ☐ 2 tsp. baking powder (well sifted)
- ☐ ¼ cup light brown sugar (optional)
- ☐ 1½-2 cups berries — fresh are best, frozen are okay (May also use: other berries, apples, peaches, or pears)

Directions:

1. Preheat oven to 350°. In a large mixing bowl, lightly beat eggs, then add milk and honey. Stir until well combined.
2. Add oats, baking powder, and seasoning, stirring until well mixed.
3. Pour mixture into well-greased 9x11 baking dish and even it out.
4. Add berries or fruit to the top of the mixture, pressing them slightly into the oatmeal mixture.
5. Optional: Sprinkle brown sugar on top.
6. Bake at 350° for about 20 minutes or until the oats are tender. Serve hot.
7. Optional (but so yummy): This baked oatmeal is delicious served with a small dab of unsweetened whipped cream on top.

As part of Lesson 21 of *Math Level 2,* there is a wonderful recipe for Saltwater Taffy. This is the very best recipe I have ever used for this treat.

Saltwater Taffy (Makes about 1 pound)

Ingredients:

- ☐ 1 cup sugar
- ☐ ¾ cup light corn syrup
- ☐ ⅔ cup water
- ☐ 1 Tbsp. cornstarch
- ☐ 2 Tbsp. butter
- ☐ 1 tsp. salt
- ☐ 2 tsp. vanilla

Directions:

1. Butter an 8x8-inch square pan.
2. In a two-quart saucepan, combine sugar, corn syrup, water, cornstarch, butter, and salt.
3. Cook over medium heat, stirring constantly, to 256° on a candy thermometer (or until small amount of mixture dropped into very cold water forms a hard ball).
4. Remove from heat; stir in vanilla.
5. Pour into pan.
6. When just cool enough to handle, pull taffy until satiny, light in color, and stiff. (If taffy becomes sticky, butter hands lightly.)
7. With scissors, cut strips into one-inch pieces.
8. Wrap pieces individually in wax paper. Candy must be wrapped to hold shape.

Lesson 24 of *Math Level 2* also contains two Peruvian recipes.

Papas a la Huancaina (Potatoes with Cheese)

Ingredients:

- ☐ 8 potatoes, peeled and cubed
- ☐ Water
- ☐ 1½ cups heavy cream
- ☐ ½ tsp. turmeric
- ☐ 3 cups Monterey Jack cheese

Directions:

1. Boil the potatoes, covered, until tender. Drain and set aside.
2. In a saucepan, heat cream over low heat. Do not bring to a boil.
3. Stir in cheese and turmeric.
4. Continue to stir until cheese is melted.
5. Add potatoes, cooking until potatoes are heated through.
6. Serve warm or cold.

Alfajores (Caramel-filled cookies)

Ingredients:

- ☐ 2 cups cornstarch
- ☐ 1 cup flour
- ☐ 1 cup sugar
- ☐ ½ tsp. baking powder
- ☐ ¾ cup butter, room temperature
- ☐ 2 eggs
- ☐ 1 tsp. vanilla
- ☐ 3 Tbsp. milk
- ☐ 1 can (13.4 oz) Dulce de Leche*
- ☐ Powdered sugar

Directions:

1. Preheat oven to 300°.
2. Combine dry ingredients in a large bowl.
3. Cut butter into small slices, drop onto dry mix and stir until mixture resembles coarse crumbs.
4. Add eggs, vanilla, and milk.
5. Knead until smooth. Let dough rest for 20 minutes.
6. Roll dough out to about ¼-inch thickness. Cut out cookies with a cookie cutter.
7. Bake for 20 minutes or until cookies begin to brown.
8. Remove from oven and cool.
9. Spread dulce de leche on one side of a cookie and top with another cookie. Roll cookie sandwich in powdered sugar. Repeat with remaining cookies.

*Dulce de Leche directions:

1. Remove label from can. Pierce the top with two holes using a can opener.
2. Place in a pot, pierced end up, and fill the pot with water to about ¼ inch from the top of can.
3. Bring to a boil. Reduce heat and simmer, uncovered, for 3 hours. You may need to add more water to the pot as it evaporates.
4. Remove from water and cool.

From Math Level 3

In Lesson 17, a Peruvian dessert reminds the twins of Grandma's blueberry bread pudding.

Throughout the years, I have occasionally made this simple dessert for my family. You can use whatever type of berry you like, or if you prefer, apples, peaches, or pears work well.

Grandma's Bread Pudding with Fruit

Ingredients:

- ☐ 6-8 slices of bread (leave them out to dry a bit) (You may also use sprouted wheat bread or gluten-free.)
- ☐ 2 Tbsp. melted butter
- ☐ 3-4 eggs (4 if they are smaller), beaten
- ☐ 2 cups milk
- ☐ ¾ cups sugar
- ☐ 1 tsp. cinnamon
- ☐ 1 tsp. vanilla
- ☐ 2 cups of berries or fruit
- ☐ Optional topping: ⅔ cup powdered sugar, 1 or 2 Tbsp. milk (stir in one at a time until the right consistency), 1 tsp. vanilla or almond extract

Directions:

1. Cut bread into cubes and place them into a greased 9x11 baking dish.
2. Combine all wet ingredients and seasonings and stir well.
3. Pour over the bread. Lightly push down on the bread to make sure it is absorbing the liquid.
4. Evenly distribute berries or fruit throughout bread cubes.
5. Bake at 350° until golden

Directions for topping:

Mix all ingredients until smooth and creamy.

Pour over slightly cooled bread pudding.

Enjoy!

Picarones (Pumpkin Fritters) this recipe is included at the end of Lesson 36 of *Math Level 3*. (Makes 12)

Ingredients:

- ☐ 1 package dry yeast
- ☐ ¼ cup lukewarm water
- ☐ 2 Tbsp. sugar
- ☐ 1 egg, lightly beaten
- ☐ 1 can (16 oz) pumpkin
- ☐ ½ tsp. salt
- ☐ 4 cups flour
- ☐ Oil, for frying
- ☐ Maple syrup

Directions:

1. In a large bowl, sprinkle the yeast over the lukewarm water and stir to dissolve.
2. Add the sugar, egg, pumpkin, and salt; combine thoroughly.
3. Add the flour, ½ cup at a time, until the dough becomes too stiff to beat with a wooden spoon.
4. Turn the dough out onto a lightly floured board and knead in enough of the remaining flour to prevent the dough from sticking to your fingers.
5. Continue kneading until the dough is smooth and elastic (about 8 minutes).
6. Shape it into a ball and place in a greased bowl. Cover and let rise in a warm place for 1 hour, or until doubled in size.
7. Punch down the dough and tear off pieces, shaping into doughnut-like rings, about 3 inches in diameter.
8. Heat about 1-inch of oil in a deep skillet and fry the fritters for about 5 minutes, turning them once, until crisp and golden brown.
9. Drain on paper towels and serve immediately with warm maple syrup.

From Math Level 4

On Lesson 6 of *Math Level 4*, Grandma has made flatbread pizza for the girls for lunch. I do not include a recipe for this in the book, but since this is such a fun and really rather easy project for kids this age, I thought I would give you my favorite recipe for flatbread here.

Garlic-Herb Flatbread (Serves 6)

Ingredients:

- ☐ 1 packet active dry yeast (this equals 2¼ tsp.)
- ☐ 2 cloves of garlic, minced (equals 1 Tbsp.)
- ☐ 1 Tbsp. combined rosemary and thyme
- ☐ ¾ tsp. sea salt or regular salt
- ☐ ½ tsp. sugar or your favorite sugar replacement
- ☐ 1¾ cups Bob's Red Mill Spelt Flour™ or other all-purpose flour
- ☐ 1 Tbsp. olive oil (plus a little more set aside for greasing the bowl)
- ☐ ¾ cups warm water (about 110°)

Instructions:

1. In a large mixing bowl, combine the yeast, garlic, herbs, salt, sugar, and flour. Whisk.
2. Make a "well" in the dry ingredients and add olive oil and ½ cup warm water to start. Stir with a wooden spoon. Add more water as needed until a dough forms.
3. Transfer to a clean, well-floured surface and knead until smooth and elastic (about 2 minutes) adding flour as needed to keep it from sticking.
4. Grease a mixing bowl with a little oil and place dough in it, rolling it around to coat it with the oil, placing it smooth side up.
5. Cover the bowl with slightly damp towel and set it in a warm place for one hour.
6. When the dough has risen (about double its size), cut it into 6 equal pieces. Arrange on a clean surface and let rest, covered with the damp towel.
7. While the dough is resting, heat a large skillet or griddle to medium-high (375°).
8. Roll out each piece of dough, one at a time (about ⅛ inch thick).
9. Lightly grease heated skillet and place rolled-out dough into a pan. Don't touch the dough for 2 to 3 minutes, then flip and cook 2 to 3 minutes on the other side.
10. Add oil between placing each piece of rolled out dough on the griddle or skillet.
11. Serve warm or make the pizza — recipe following this one.

Pizza Toppings for Flatbread Pizza:

These are a fun way to create personal-sized pizzas. Makes 6.

☐ Your choice of marinara sauce

☐ Any type of meat you would like

☐ Shredded cheese: mozzarella, colby jack, cheddar, or munster work the best

☐ Sautéed or canned mushrooms

Instructions:

1. Each person creates their flatbread pizza.
2. Place them on baking sheets and bake at 400° until the cheese is melted and the meat is heated through.

In Lesson 21, there are two snack recipes. These are all-time favorites in my family; my kids have made them so often, they no longer have to look at a recipe. The smoothie pops also make delicious non-frozen smoothies.

Orange-Banana Smoothie Pops

Ingredients:

☐ 1 container (7 oz) Greek yogurt

☐ ⅔ cup thawed orange juice concentrate

☐ 2½ large bananas (for a thick, non-frozen smoothie, use frozen bananas)

☐ Zest of 1 lime

☐ 1 Tbsp. fresh lime juice

Instructions:

1. Puree Greek yogurt, thawed orange juice concentrate, bananas, lime zest, and fresh lime juice in a blender. (If drinking as a non-frozen smoothie, pour into chilled glasses and enjoy!)
2. Pour into six paper cups and poke a craft stick into the center of each. Freeze until smoothie pops are solid (about 4 hours).
3. To release pops, peel away the paper. Enjoy!

Tasty Trail Mix — Makes 8-12 servings

Ingredients:

- ☐ 1¾ cups dried fruit (such as apricots, apples, pears, bananas, pineapple, and dried cranberries or blueberries)
- ☐ ⅔ cup raisins
- ☐ 1½ cups salted sunflower seeds
- ☐ 1½ cups nuts of your choice
- ☐ 1¼ cups of chocolate chips

Instructions:

Mix all together and enjoy!

Fresh Tomato Salsa (Pico de Gallo)

In Lesson 28 of *Math Level 4* is a recipe for Fresh Tomato Salsa. This pico de gallo recipe is so delicious that I try to keep a container of it in my refrigerator at all times during the summer months when the tomatoes are fresh and plentiful.

Ingredients:

- ☐ 6 tomatoes, chopped
- ☐ 1 cup finely diced onion
- ☐ 3 or 4 jalapeño peppers finely chopped
- ☐ 1 green bell pepper finely chopped
- ☐ ¾ cup chopped fresh cilantro
- ☐ 2 fresh garlic cloves
- ☐ 1 Tbsp. of lime juice
- ☐ Salt to taste

Instructions:

Mix all ingredients together and chill for one hour. Enjoy with tortilla chips.

Included in Lesson 5 of *Math Level 5* is my favorite cookie recipe. I created this recipe when I was pregnant with our 4th child. I was craving sweet and healthy treats, and this one was the best one I concocted. So yummy. So filling. So healthy.

Orange Zesty Oatmeal Raisin Cookies

Ingredients:

- ☐ ¾ cup raisins
- ☐ ½ cup fresh orange juice
- ☐ 4 Tbsp. orange juice concentrate
- ☐ ⅓ cup butter (softened) or ¼ cup vegetable oil
- ☐ 2 tsp. vanilla
- ☐ 1 Tbsp. orange zest
- ☐ 1 tsp. lemon zest
- ☐ 3 Tbsp. molasses
- ☐ 2 eggs
- ☐ 1½ cup quick-cooking oats
- ☐ 1½ cup all-purpose flour
- ☐ 2 tsp. baking powder
- ☐ ¼ tsp. salt
- ☐ ¼ tsp. cinnamon
- ☐ ¼ tsp. ginger
- ☐ ½ cup chopped pecans (optional)

Directions:

1. In a small saucepan, bring orange juice to a boil.
2. Add the raisins, cover, and remove from the heat. Allow to sit for 15 minutes, drain the juice off of raisins.
3. Cream together: oil (or butter), vanilla, and zests.
4. Add egg, juice concentrate, and molasses and beat until well blended.
5. Sift together dry ingredients.
6. Stir oats and drained raisins into wet mixture and add dry ingredient mixture slowly, mixing well.
7. Stir in nuts (optional).
8. Refrigerate batter for at least 2 hours.
9. Preheat oven to 350°. Spoon tablespoons of dough onto parchment-lined baking sheets and flatten slightly with a spoon.
10. Bake at 350° for 10-12 minutes. Cookies should be firm but soft.

Lesson 16 of *Math Level 5* includes: **Grandma Violet's Cranberry Christmas Punch Delight**

Ingredients:

- ☐ ⅓ cup white pure cane sugar
- ☐ 2 cups cranberry juice
- ☐ 3 Tbsp. almond or vanilla extract
- ☐ 1 (2-liter) bottle of ginger ale

Directions:

1. Mix first three ingredients and refrigerate for 24 hours.
2. Add ginger ale before serving.
3. Be aware! This recipe is high in sugar, so enjoy sparingly.

Lesson 19 of *Math Level 5* contains a recipe for:

Grandma Violet's Chicken Veggie Soup (Serves 8-10)

Instructions:

1. Boil 4 or 5 bone-in, but skinless, chicken breasts in about 5 quarts of water. The water will boil down, and the broth of the chicken will thicken it.
2. Boil chicken until it is falling off of the bone.
3. Take chicken out of the broth and let cool.
4. Pick chicken off of bones and set aside to completely cool. Chop as fine as you would like.
5. Chop and prepare any veggies that you want to put into your soup. If you want to go easy, get a bag of mixed frozen veggies (our favorite is green bean, corn, peas, carrots).
6. Add another quart of chicken broth (homemade* or store-bought).
7. Bring the pot of broth to a boil and add the veggies.
8. Add the chicken.
9. Season with salt, a little black pepper, and a bay leaf (remove bay leaf before eating).
10. Let soup simmer for a while.
11. Add 1½ cups of rice or 2 cups of egg noodles and cook.

* To make homemade, simply add water to the pot and use chicken soup base to taste. It will take 1/4 cup of soup base, but start with about half that amount and add more until it tastes right.

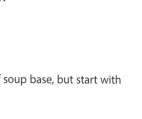

In Lesson 23, your student will be creating their own Mexican Fiesta! Here is a delicious one-pan meal.

Cheesy Chicken Enchilada Bake

Ingredients:

- ☐ 1 pound boneless, skinless chicken breast (about 2 cups)
- ☐ ½ cup water
- ☐ 1 Tbsp. chili powder
- ☐ 1 can (15 oz) low sodium black beans, rinsed and drained
- ☐ 1 cup frozen corn
- ☐ 1 cup salsa
- ☐ 8 whole-wheat tortillas
- ☐ Cooking spray
- ☐ ½ cup 2%-fat shredded cheddar cheese

Directions:

1. Cut chicken breast into 4-5 chunks. Simmer in a large saucepan with water and chili powder. Cook until internal temperature is 165° (about 10 minutes).

2. Remove chicken from pan. Cut or shred into small chunks and return to pan. Add beans, corn, and salsa to saucepan. Cook until hot, about 2 minutes. Remove from heat.

3. Spread ½ cup of chicken mixture down the center of each tortilla. Roll up and place seam-side down in greased 9×13 pan.

4. Spread any leftover chicken mixture over the top of the enchiladas.

5. Bake at 375° for 12-15 minutes.

6. Sprinkle cheese on top of the enchiladas during the last 5 minutes of cooking.

7. Serve immediately.

From Math Level 6

In Lesson 9 of *Math Level 6* is my family's Favorite Gluten-Free Chocolate Chip Cookies recipe.

I include it here along with the directions for making them.

Ingredients:

- ☐ ½ cup butter (room temperature)
- ☐ 2 eggs
- ☐ 2 Tbsp. vegetable oil
- ☐ 1½ tsp. vanilla
- ☐ 1 cup packed brown sugar
- ☐ 2½ cups of gluten-free flour mix (Pillsbury Best Multi-Purpose Gluten-Free Flour Blend™ is our favorite for baking)
- ☐ 2 tsp. gluten-free baking powder (well sifted to remove any clumps)
- ☐ 1 tsp. baking soda (well sifted to remove any clumps)
- ☐ ½ tsp. salt
- ☐ 1¾ cups chocolate chips
- ☐ Optional: ⅔ cup chopped nuts

Directions:

1. In a large mixing bowl (a stand mixer works best, but a hand mixer will suffice), beat eggs, then add the brown sugar. Cream well. Add butter, vanilla, and oil, beating well between each ingredient. Scrape the bowl and mix again.

2. In a separate bowl, combine flour, baking powder and soda, and salt.

3. Turn the mixer on low and slowly add dry ingredients to wet ingredients, one cup at a time. Scrape sides of mixing bowl as needed.

4. Mix the dough well, but do not over-beat.

5. Fold in chocolate chips and nuts. Preheat oven to 350°.

6. Use two spoons to drop even amounts of dough onto parchment paper-lined cookie sheets.

7. Bake at 350° until cookies are golden on the bottom (about 10-12 minutes).

8. Remove from oven and allow to cool slightly before transferring to a paper towel to cool completely.

In Lesson 11 of *Math Level 6*, I talk about Grandma's Chicken Veggie soup. I have included the recipe for this yummy soup in the recipe section for *Math Level 5* in this *Companion*.

Section #3: Math Games

Odds and Evens Games

One, Two, Three, GO!

- ☐ **Number of players:** 2 to 6
- ☐ **Age:** any student learning and practicing the concept of odd and even numbers.
- ☐ Needed materials: just your hands
- ☐ This game is simple and fun — think Rock, Paper, Scissors with a twist…

Instructions:

1. Make two teams – one odd and one even. (Each team is one person if playing with two players.)
2. Players sit facing each other.
3. Players say: "one, two, three, GO!" (in unison) while they smack their closed fist on top of their open hand. When they reach "GO!", they stick out one or two fingers from their closed fist.
4. Count the total amount of fingers sticking out on all players. If it's odd, the odd team wins; if it's even, the even team wins.

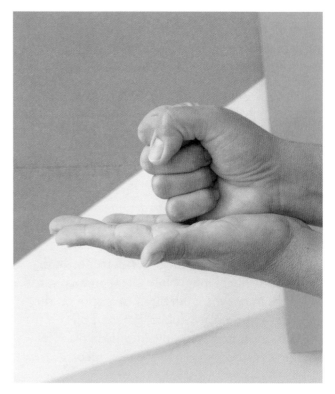

Roll 'Em!

- ☐ **Number of players:** 2 to 8
- ☐ **Age:** any student learning and practicing the concepts of odd and even numbers.
- ☐ Needed materials: pair of dice and a 3-minute timer

Instructions:

1. Make two teams – one odd and one even (each team is one person if playing with two players).
2. One player is assigned as the scorekeeper. Job: write a column of numbers rolled for each team.
3. Players gather in a circle, with the teams sitting intermingled (if more than 3 or four players).
4. Set the timer.

5. Go clockwise around the circle, with each player rolling the dice, adding the numbers and deciding if the sum is odd or even. The scorekeeper writes it down in the correct column.
6. At the end of the 3 minutes, whichever team has more numbers in their column wins.

Place Value Games/Activities

Build-a-number Kit

Age: any student learning place value concepts.

Note: This activity/game takes a bit of prep time but is well worth the investment. Once you have the game prepared, store the pieces in a large Ziploc bag, and use it once a week or so. This kit may be kept and used throughout all of your child's elementary years.

Needed materials:

☐ A large package of white card stock, scissors or paper slicer, a ruler, nine colors of markers

Instructions for preparations:

1. Begin with 60 strips of paper, measuring 2½ inches wide and the following lengths: (10) 1 inch long, (10) 3 inches long, (10) 5 inches long, (10) 7 inches long, (10) 9 inches long, (10) 11 inches long. Note: the 0 on each card needs to be written in the correct spot on the strips to be a place holder. For example: on the 100's strip, write it in the hundred's place, on the 10's strip, write it in the tens' place.

2. On the 1-inch-long strips, write (one number on each card): 0, 1, 2, 3, 4, 5, 6, 7, 8, 9 (all in one color)

3. On the 3-inch-long strips, write (one number on each card): 0, 10, 20, 30, 40, 50, 60, 70, 80, 90 (all in one color not used before)

4. On the 5-inch-long strips, write (one number on each card): 0, 100, 200, 300, 400, 500, 600, 700, 800, 900 (all in one color not used before)

5. On the 7-inch-long strips, write (one number on each card): 0, 1,000, 2,000, 3,000, 4,000, 5,000, 6,000, 7,000, 8,000, 9,000 (all in one color not used before)

6. On the 9-inch-long strips, write (one number on each card): 0, 10,000, 20,000, 30,000, 40,000, 50,000, 60,000, 70,000, 80,000, 90,000 (all in one color not used before)

7. On the 11-inch-long strips, write (one number on each card): 0, 100,000, 200,000, 300,000, 400,000, 500,000, 600,000, 700,000, 800,000, 900,000 (all in one color not used before)

You can add on to this kit by creating 10 each of these: 13-inch strip, 15-inch strip, 17-inch strip.

8. On the 13-inch strips, write: 0 (in the one million's place) 1,000,000, 2,000,000, 3,000,000, 4,000,000, 5,000,000, 6,000,000, 7,000,000, 8,000,000, 9,000,000 (all in one color not used before)

9. On the 15-inch strips, write: 0 (in the ten million's place) 10,000,000, 20,000,000, 30,000,000, 40,000,000, 50,000,000, 60,000,000, 70,000,000, 80,000,000, 90,000,000 (all in one color not used before)

10. On the 17-inch strips, write: 0 (in the hundred million's place) 100,000,000, 200,000,000, 300,000,000, 400,000,000, 500,000,000, 600,000,000, 700,000,000, 800,000,000, 900,000,000 (all in one color not used before)

Directions for use:

Have your student create and read numbers by stacking the strips of paper on top of each other. Have them explain what they are doing.
For example: tell your student to create the number 37,905,941. You may write it on the whiteboard for them to see it.

They would stack:

1. The 30,000,000 strip
2. The 7,000,000 strip
3. The 900,000 strip
4. The zero place holder strip for the 10,000's strip
5. The 5,000 strip
6. The 900 strip
7. The 40 strip
8. The 1 strip

Number Toss

Age: any student learning advanced place value concepts.

Needed materials:

- ☐ A large piece of poster board
- ☐ A die
- ☐ A whiteboard or large piece of paper

Setup:

1. Create a large floor playing mat by following these directions:

2. Divide the poster board into three rows of three (9 squares in all).

3. In each square write: 1's, 10's, 100's, 1,000's, 10,000's, 100,000's, 1,000,000's, 10,000,000's, 100,000,000's.

To play:

1. Have the students take turns building a number by following these directions:

2. Lay the mat on the floor.

3. The student tosses the die into each of the squares and writes down what the number would be using the value on the die. After the student has tossed the die 9 times (one for each square), they "build" their number by placing the numbers gathered into the correct order.

4. You can make this into a two-player game by seeing who can come up with the biggest number.

For example:

1. The die landed on 4 in the 100,000's square, they would write down 400,000.

2. The die landed on 1 in the 10's square, they would write down 10.

3. The die landed on 9 in the 10,000's square, they would write down 90,000.

4. The die landed on 5 in the 100,000,000's square, they would write down 500,000,000.

5. The die landed on 2 in the 1's square, they would write down 2.

6. The die landed on 6 in the 100's square, they would write down 600.

7. The die landed on 3 in the 1,000's square, they would write down 3,000.

8. The die landed on 2 in the 10,000,000's square, they would write down 20,000,000.

9. The die landed on 9 in the 1,000,000's square, they would write down 9,000,000.

After they are finished gathering their numbers, they would place them in order: 529,493,612.

Recitation bee

Long ago, in centuries past, blab-school teachers knew the value of calling on students for recitations. Although there were times when these one-room schoolhouses had a few moments of silence, much of the time, there was some type of noise in the form of groups of children being called upon to give an oral narration on something they were learning. With the coming of large city schools and crowded classrooms — all being watched over by one teacher — came the use of fill-in-the-blanks and silent children sitting in rows. Since there was no way to use the old-fashioned practice of oral narration and recitations with that many students, teachers were no longer trained to teach that way. Textbooks and workbooks replaced the living books and slates used for hundreds of years before that. Teachers are now instructed to teach the curriculum, not the individual student.

In our homeschools, we have no reason to depend on the "standardized curriculum," which uses extensive testing to teach our children. We can, instead, actually interact with our children to ascertain what they are learning. One of the ways we can do this is through oral narration and recitation.

The main reason we homeschool is to build relationships with our children. We, as parents are, day by day, moment by moment, building a safe base for our children's future launch. Part of building this safe base comes in the form of teaching communication. Fluid communication comes from fluid thought. Fluid thought comes from strong reasoning skills. Strong reasoning skills do not just happen. They take work and effort.

In our home, I used recitation bees as a way to "test" my children in a fun but competitive way. Once a month, I wrote into my planner a day that would be our special recitation day. I planned ahead for each of my kids, making a list of what they were going to be expected to recite at the end of the month. I gave a list to each of them as well. Included in this list were aspects of every "subject" that they were studying. The older students would be called on to give an oral presentation (using props and timeline when necessary) about: a certain history period, event, or person, an argument presentation for what they were learning in apologetics or science, a recitation of poetry or Scriptures they had been memorizing, and an overview of certain math or personal finance concepts.

My younger students did show-and-tells (like the ones worked into the math series), poetry, Scripture, and math fact recitations. They also did show-and-tells for what they were learning in geography, history, and science. On these special days, we would end the day by watching a movie and having a special snack. The kids knew that these presentation days would count for about 50% of their overall "grade" in that subject. They knew they were expected to be able to fluently and accurately give their recitation without fumbling and mumbling. There were times when one of them wasn't prepared, and they didn't pass that month's worth of schooling.

In many ways, I consider myself to be a pretty relaxed homeschool mom, but in other ways, I have high expectations. The ability to deliver a recitation is a major place of high expectation for me. I know that God has blessed each of my children with a good brain (even the ones who struggle in certain ways), and I expect them to take full responsibility for their part in developing themselves to their full potential. Memorization and recitation have been proven to help develop cognitive ability and maturity levels in young people.

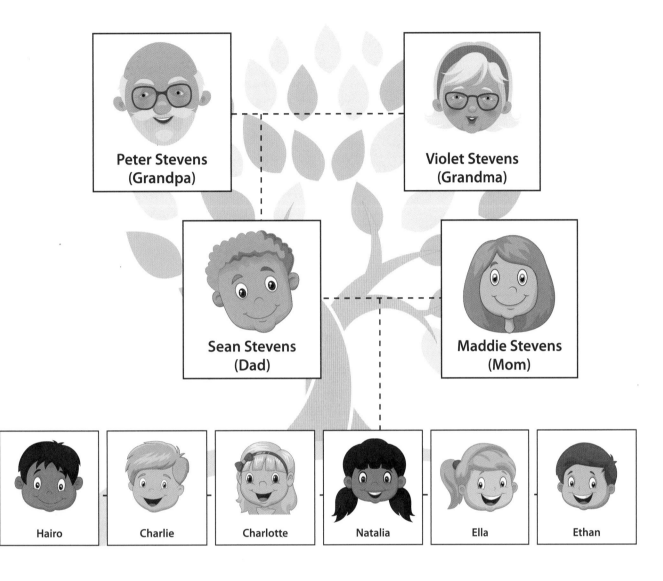

Other family members mentioned:

In *Math Level 4*: Great Grandpa Sam (Grandpa Peter's father who was killed in World War II when Peter was a baby), Great Grandpa Fred (Grandpa Peter's stepfather who married Peter's mother when he was 3 years old, also a World War II veteran, and the only father Peter knew).

In *Math Level 5*: Maddie Stevens' sister, Kate, and her husband, Justin, and their children, Sean, Abby, and Danielle.

Section #5: Synopsis of Storyline

Math Level K:

Meet Charlie and Charlotte, five-year-old twins who live with their parents in a small town in rural Minnesota. The twins love to explore the world around them and, along with their parents and grandparents, discover that life as kindergarteners is full of all kinds of new and exciting adventures. In this age-appropriate prequel to the storyline in *Math Levels 1- 6*, your student is drawn into the twins' learning adventures. Clap, skip, hop, and climb your way through *Math Level K*, which perfectly coordinates with *Simply K*, the Kindergarten curriculum from Master Books!

Math Level 1:

The twins, Charlie and Charlotte, are spending part of the summer with their grandparents out on the farm, while their parents work in Peru to build a childrens' home. Throughout their time with their grandparents, the twins learn many lessons about the life cycles around them in nature. They also learn about counting, place value, telling time, and writing numbers. They help with the chores around the farm and learn how numbers and patterns are used in everyday life. By the end of their visit with their grandparents, the twins find out that their mom is expecting a new baby sister and that their parents have decided to homeschool them on their return home. Character training and Bible stories are woven throughout the story as Grandpa Peter and Grandma Violet teach the children how the attributes of God are displayed all around us in nature and in our family relationships.

Math Level 2:

In *Math Level 2*, the children have returned home and have begun their homeschooling journey. The storyline for this level is centered around family relationships, hands-on learning, the arrival of their new baby sister, Ella, and learning more about the world in which they live. In *Level 2*, the student will be drawn into a new friendship that has developed between the twins and a set of siblings that live in the Peruvian Children's Home which their parents helped build during *Math Level 1*'s storyline. The twins interact with the Peruvian children, Hairo and Natalia, writing them letters, buying them presents, and helping to raise money to help support them.

Math Level 3:

Math Level 3's storyline is centered around a family missions trip to work in the Peruvian Children's Home, where Hairo and Natalia live. The summer is spent playing and spending time with their friends, all the while trying to figure out a way to find a new mom and dad for Hairo and Natalia. Exciting and sometimes scary events, such as a bad storm which knocks out the electricity and destroys people's homes in Lima, mark the family's stay in Peru. Your student will be swept up in the real-life experiences and drama as Charlie and Charlotte step outside of their comfort zone to help those around them. Math concepts are woven throughout the story as the children learn and discover the strange new world of South America. The story culminates as the twins pray for a miracle that would bring their dear friends a new family. What will happen? Will Hairo and Natty find a new family?

Math Level 4:

The storyline for *Math Level 4* is full of all kinds of fun and exciting experiences as the family takes a vacation across the upper Midwest states of the US. Your student will learn and practice new math concepts as they travel with Charlie, Charlotte, Hairo, and Natty. They will discover Mount Rushmore, the Black Hills, and many other historical and geographical marvels. They will continue to build their character and discover the amazing attributes of God displayed in creation around them. As the children in the story grow and develop, they share their experiences with your student, and in the midst of it all, reassure them that they are not the only ones who are growing up and dealing with changing emotions and feelings! The storyline for *Math Level 4* takes a step up in maturity as it adjusts for the ages of all of the children involved.

Math Level 5:

Math Level 5's storyline is a special treasure of surprises and activities. Your student will enjoy the everyday life fun of the Stevens children, learning alongside them as they discover concepts such as being rewarded through pay for hard work. They will travel with the family to spend a few weeks at a camp where the children's parents are teaching a survival course to the campers. They will also gain more responsibility and freedom as they follow the directions to create their very own Mexican Fiesta — decorations and all! *Level 5*'s storyline was carefully crafted to speak to the hearts and minds of children this age, gently addressing some of the major life issues we all face. Math concepts are built upon, character lessons are woven throughout, and relationships are created and strengthened.

Math Level 6:

The storyline of *Math Level 6* is a special gift from me to your student. Before creating this level, I prayed for direction in a very specific way since I knew that young people this age may have a hard time accepting a storyline as part of their math. I certainly did not want them to feel like this concept was babyish or condescending in any way, shape, or form. I wanted them to feel invited into this volume in a deliberate way. The storyline for this level is unique in that I did not try to build a story that would be math-oriented in any way. If there were a math concept that I could include, I would include it, but it wasn't my focus. Instead, I wanted to address life issues that young people this age commonly face. I wanted to talk to them about loss, death (or near-death), uncertainty, major life changes, bullying, allowing God to use our pain, and living to please Him instead of ourselves. I wanted to bring the students in this age bracket a cohesive storyline and a thorough review of all of the math concepts that they have learned so far, as well as taking those concepts much deeper. I wanted to make sure your student had taken the time to really truly own their education and understand that it is up to them to continue learning.

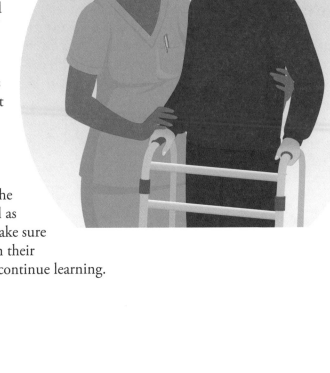

Section #6: *Math K* and Readiness Tests and Solutions for *Levels 1-6*

Dear Parent,

When placing your child in *Math K*, please use the following lists as a guide in your decision.

The first section outlines a few points in cognitive and developmental readiness. This is truly the most important section. If your child's growth has not yet reached these developmental markers, please wait until they do. Remember, all children develop at their own speed, and no two children are alike.

The second list is academically focused. In this section, if your student can accomplish any three of the six, *and* they have reached the developmental markers in the first section, *Math K* will fit well.

—Angela O'Dell

The following is a section of information from *Math K* co-author, Carrie Bailey.

Is your child ready cognitively and developmentally*?

- Readiness doesn't mean just knowing the academic basics. It means a child has a willing attitude and confidence in the process of learning. Does your child display a good attitude and a healthy state of mind?

- Developmentally, children of this age cannot learn on demand. They learn because they want to. Does your child show an interest in learning and exploring?

- Play must be taken seriously. Play is a child's work. Does your child willingly engage with you when you suggest structured playtime for at least a few minutes at a time?

* All information has been cross-referenced with the National Association of Education for Young Children [NAEYC], but neither this teaching companion or the Math for a Living Education series are based on Common Core.

Is your child ready academically?

1. Do they notice differences in their daily lives, such as night and day?
2. Do they have pencil control to trace lines?
3. Can they cut a basic line?
4. Are they familiar with a square, circle, and triangle (can they point out the differences between the shapes)?
5. Can they put a simple puzzle together?
6. Does your child know the numbers 1, 2, 3, and count three objects?

Math Lessons for a Living Education teaches math in a unique way—these readiness tests will guide you in determining the best level in which to place your student.

Each test contains the skills and concepts your student must know and understand in order to enter that level. These are the prerequisite skills your child must understand before beginning.

As your student works through these tests, make sure they understand:

- How each process is performed
- Why each process works

As your student completes each problem, ask them to show or tell you what they are doing and why they are doing it. Future success in mathematics relies upon your student understanding both the why and how of math.

Example readiness scenarios:

If your student can . . .

- Easily pass the test for *Math Level 3* and understands both why and how they utilize those mathematical concepts, but struggles in the *Math Level 4* test, your student is ready to begin *Math Level 3*.

- If your student can pass the test for *Math Level 5*—but cannot show or tell you how concepts work (they know how to "fill in the blanks"), your student should begin *Math Level 4* in order to fill in learning gaps and create a true understanding of the concepts.

- If your student can pass the test for *Math Level 4*, but has one or two learning gaps (they are still a little shaky on a topic or two), you may evaluate the topics covered in both *Math Level 3* and *4* and use your discretion in placing them. We would recommend working through *Math Level 3* at an accelerated pace; however, you may choose to place your student in *Math Level 4* and fill in learning gaps together.

Level 1 Readiness Test

This is a checklist to assess your student's readiness to begin *Math Lessons for a Living Education Level 1.* **If your student can accomplish all the activities in this test, they are prepared to begin *Math Level 1.***

☐   Know left from right

☐   Draw a straight line

☐   Trace a looping line

☐   Write their name, holding pencil correctly

☐   Use scissors correctly to cut lines at the bottom of this page

☐   Know colors (blue, red, yellow, orange, purple, green, brown, black, white)

☐   Follow directions successfully (i.e. play Mother May I?, giving 2-step instructions)

Level 2 Readiness Test

This placement test assesses your student's readiness to begin *Math Lessons for a Living Education Level 2*. Please discuss any missed problems with the student in order to understand the reason he or she missed them. Instructions for grading are at the beginning of each section. **If your student completes this test and understands the concepts, they are prepared to begin** *Math Lessons for a Living Education Level 2.*

Part 1: The student should make no more than 2 mistakes on each of these sections.

Section one: Teacher, instruct your student to write the numbers 0-100 on the following lines.

Part 2: Teacher, instruct your student to underline every number on the previous page that is in the ones' place with a red crayon/pencil, every number in the tens' place with a green crayon/pencil, every number in the hundreds' place with a blue crayon/pencil.

Orally, have your student answer these questions.

☐   In the number 236, what does 6 stand for?
a) six groups of ten
b) six groups of one
c) six groups of one hundred

☐   In the number 236, what does 3 stand for?
a) three groups of ten
b) three groups of one
c) three groups of one hundred

☐   In the number 236, what does 2 stand for?
a) two groups of ten
b) two groups of one
c) two groups of one hundred

Section two: The student should make no more than 1 mistake on each of these points.

Point 1: Have your student draw hands on these clocks to show the correct time.

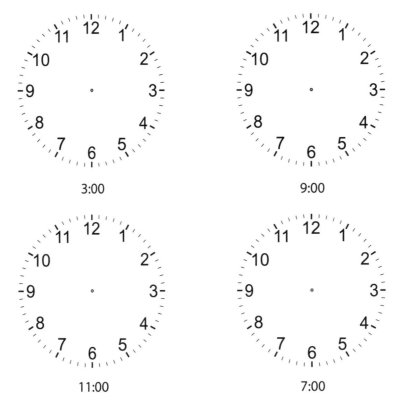

3:00   9:00

11:00   7:00

Point 2: Teacher, have your student answer these quickly. They should do these from memory; watch them carefully and take note of the ones they have to think or count to answer. (This is to see if your student understands the concept of addition. If they can answer from memory, this is a plus, but not absolutely necessary.)

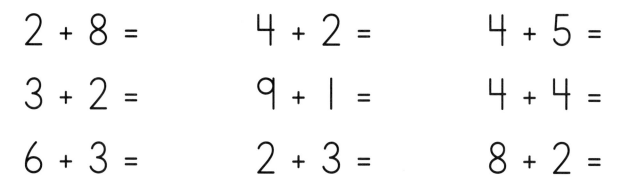

$$2 + 8 = \qquad 4 + 2 = \qquad 4 + 5 =$$

$$3 + 2 = \qquad 9 + 1 = \qquad 4 + 4 =$$

$$6 + 3 = \qquad 2 + 3 = \qquad 8 + 2 =$$

Point 3: Teacher, have your student answer these quickly. They should do these from memory; watch them carefully and take note of the ones they have to think or count to answer. (This is to see if your student understands the concept of subtraction. If they can answer from memory, this is a plus, but not absolutely necessary.)

$$10 - 2 = \qquad 8 - 3 = \qquad 6 - 2 =$$

$$9 - 7 = \qquad 10 - 5 = \qquad 9 - 5 =$$

$$10 - 8 = \qquad 7 - 4 = \qquad 6 - 5 =$$

Section three: The student should make no more than 1 mistake on each of these.
Have your student narrate to you the relationship between addition and subtraction. Do not help or coach your student at all. It extremely important that they understand the relationship between these two operations. If your student has done well on the other parts of this placement test, but does not understand this concept of relationship, please take a few minutes to use manipulatives to show them with the hands-on/visual/auditory approach. If they are not understanding this concept easily and are not able to narrate back to you as they show you with the manipulatives, they are not ready for *Math Level 2*.

Level **3** Readiness Test

This placement test assesses your student's readiness to begin *Math Lessons for a Living Education Level 3*. Please discuss any missed problems with the student in order to understand the reason that he or she missed them. Instructions for grading are at the beginning of each section. **If your student completes this test and understands the concepts, they are prepared to begin** *Math Lessons for a Living Education Level 3*.

Section one: The student should make no more than 2 mistakes on each of these points.

Point 1: Fill in the chart correctly.

	Thousands	Hundreds	Tens	Ones
6,011				
792				
4,009				
8,178				
2,060				

Point 2: Look at the numbers in the chart above. Color each even number green. Color each odd number blue.

Point 3: What numbers do odd numbers end in? _____

What numbers do even numbers end in? _____

Section two: The student should make no more than 2 mistakes on each of these points.

Point 1: Write the correct time shown on each clock.

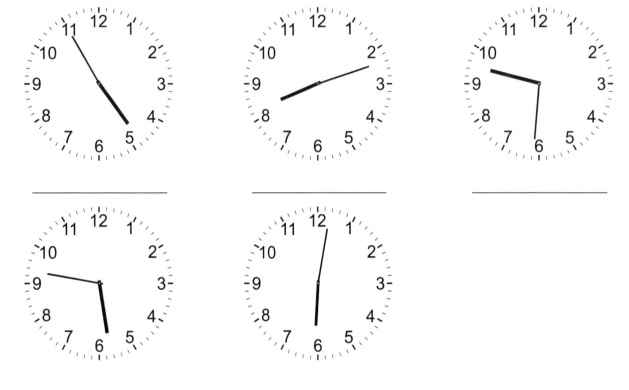

_____   _____   _____

_____   _____

Point 2: Count the money and write the correct amount.

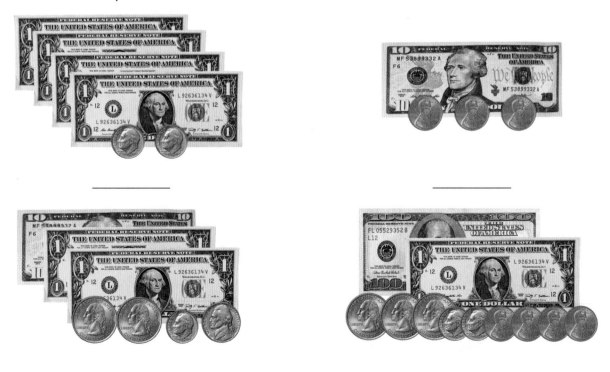

_____   _____

_____   _____

Point 3: Find the perimeter of each shape.

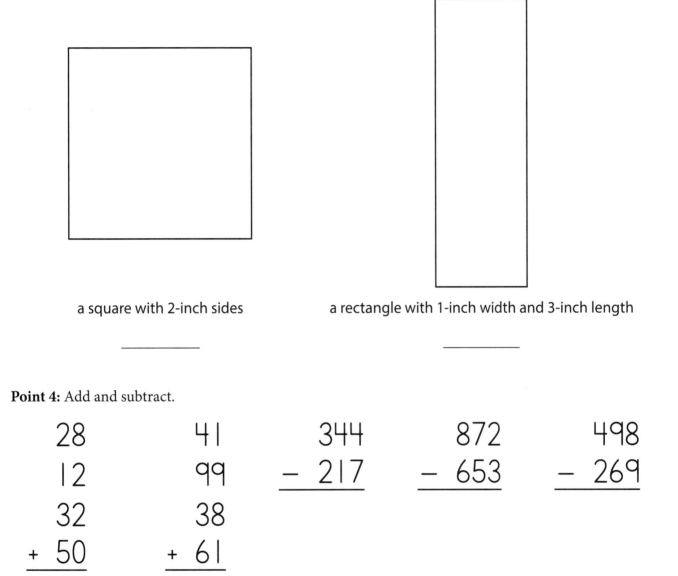

a square with 2-inch sides   a rectangle with 1-inch width and 3-inch length

_____   _____

Point 4: Add and subtract.

$$
\begin{array}{r} 28 \\ 12 \\ 32 \\ + \ 50 \\ \hline \end{array}
\qquad
\begin{array}{r} 41 \\ 99 \\ 38 \\ + \ 61 \\ \hline \end{array}
\qquad
\begin{array}{r} 344 \\ - \ 217 \\ \hline \end{array}
\qquad
\begin{array}{r} 872 \\ - \ 653 \\ \hline \end{array}
\qquad
\begin{array}{r} 498 \\ - \ 269 \\ \hline \end{array}
$$

Point 5: Measure these lines. Write the length.

☆————————————————

☆———————————

☆——————————————————————

Level 4 Readiness Test

This placement test assesses your student's readiness to begin *Math Lessons for a Living Education Level 4.* Please discuss any missed problems with the student in order to understand the reason that he or she missed them. Instructions for grading are at the beginning of each section. **If your student completes this test and understands the concepts, they are prepared to begin *Math Lessons for a Living Education Level 4.***

Section one: The student should make no more than 2 mistakes on each of these points.

Point 1: Add and subtract.

$$
\begin{array}{r} 4,561 \\ 5,198 \\ + \ 3,210 \\ \hline \end{array}
\qquad
\begin{array}{r} 3,290 \\ + \ 9,229 \\ \hline \end{array}
\qquad
\begin{array}{r} 823,197 \\ + \ 29,510 \\ \hline \end{array}
\qquad
\begin{array}{r} 329,528 \\ - \ 32,999 \\ \hline \end{array}
\qquad
\begin{array}{r} 56,291 \\ - \ 13,897 \\ \hline \end{array}
$$

Point 2: Round these numbers to the nearest 10.

23

891

466

138

Round these numbers to the nearest 100.

189

2,345

982

312

Round these numbers to the nearest 1,000.

3,780

12,428

9,621

13,289

Point 3: Complete this multiplication chart.

x	1	2	3	4	5	6	7	8	9	10
1										
2										
3										
4										
5										
6										
7										
8										
9										
10										

Point 4: Narrate to your teacher the relationship between multiplication and division. Use manipulatives to demonstrate your understanding.
(Note to the teacher: this point is a make or break. If your student does not understand multiplication and division well enough to confidently and clearly narrate to you the relationship between multiplication and division, seriously consider placing them in the previous level in this series.)

Section two: The student should make no more than 2 mistakes on each of these points.

Point 1: Find the area. Write the equations for each one.

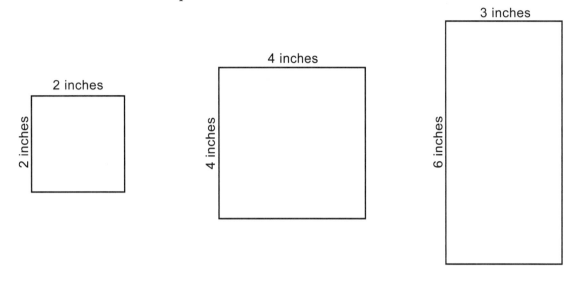

_____   _____   _____

Point 2: Correctly color each circle to show the fraction written under each one.

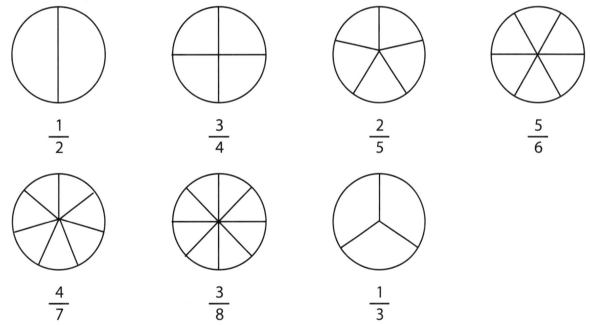

Point 3: Solve these word problems.

There were 32 tulips at the flower stand. If 4 ladies bought an equal number of the tulips, how many tulips did they each buy?

The family drove 126 miles before lunch. After lunch, they drove 253. How many more miles did they drive in the afternoon than in the morning? How many miles did they drive in the morning and the afternoon together?

Point 4: Solve these problems.

Circle groups of 3. $\frac{1}{6}$ of 18 =   6 x _____ = 18

Circle groups of 5. $\frac{1}{2}$ of 10 =   2 x ____ = 10

Circle groups of 4. $\frac{1}{3}$ of 12 =   3 x _____ = 12

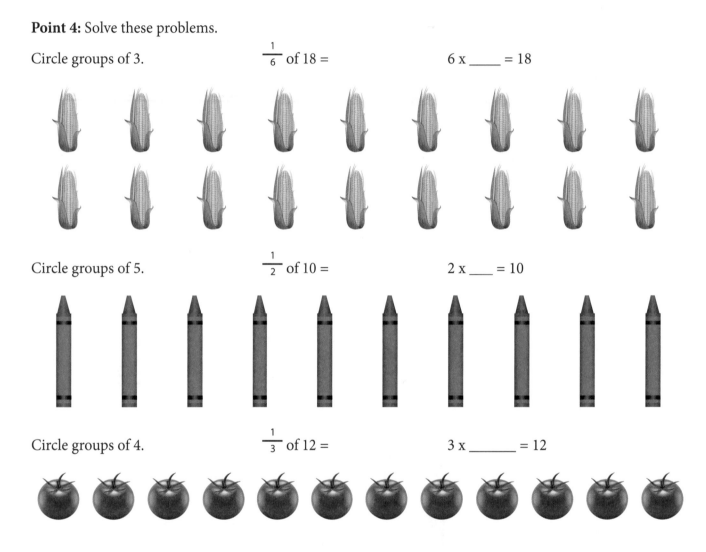

Point 5: Write the Roman numeral for each number.

I _____

5 _____

10 _____

50 _____

100 _____

1,000 _____

Level 5 Readiness Test

This placement test assesses your student's readiness to begin *Math Lessons for a Living Education Level 5*. Please discuss any missed problems with the student in order to understand the reason that he or she missed them. Instructions for grading are at the beginning of each section. **If your student completes this test and understands the concepts, they are prepared to begin** *Math Lessons for a Living Education Level 5.*

Section one: The student should make no more than 2 mistakes on each of these points.

Point 1: Add and subtract.

$$
\begin{array}{r}
289,591 \\
429,398 \\
+ \ 129,510 \\
\hline
\end{array}
\qquad
\begin{array}{r}
87,109,792 \\
+ \ 1,349,029 \\
\hline
\end{array}
\qquad
\begin{array}{r}
890,573 \\
+ \ 449,977 \\
\hline
\end{array}
$$

$$
\begin{array}{r}
23,369,219 \\
- \ 57,259 \\
\hline
\end{array}
\qquad
\begin{array}{r}
566,773 \\
- \ 233,783 \\
\hline
\end{array}
$$

Point 2: Multiply.

$$
\begin{array}{r}
45 \\
\times \ 33 \\
\hline
\end{array}
\qquad
\begin{array}{r}
85 \\
\times \ 41 \\
\hline
\end{array}
\qquad
\begin{array}{r}
93 \\
\times \ 55 \\
\hline
\end{array}
\qquad
\begin{array}{r}
72 \\
\times \ 29 \\
\hline
\end{array}
\qquad
\begin{array}{r}
25 \\
\times \ 12 \\
\hline
\end{array}
$$

Divide.

$$
4\overline{)9} \qquad\qquad 3\overline{)8} \qquad\qquad 5\overline{)6}
$$

Point 3: Word Problems

The toy shop had 2,872 boomerangs in stock for the Christmas sale. After the sale, there were 1,988 boomerangs still in stock. The store decided to place half of the boomerangs on the clearance shelves, and donate the other half to a missions organization. How many boomerangs were donated to the missions organization?

When the boomerangs were delivered to the missions organization, they were equally packaged in two large boxes. How many were in each box? When the workers at the organization opened one of the boxes, they found that a dozen boomerangs had been damaged in the shipment. How many boomerangs were undamaged in that box?

Point 4: Add and subtract these fractions.

$$\frac{3}{7} + \frac{2}{7} =$$ $$2\frac{2}{5} + 1\frac{1}{5} =$$ $$3\frac{5}{9} + 2\frac{1}{9} =$$

$$\frac{3}{11} + \frac{6}{11} =$$ $$6\frac{2}{3} - 4\frac{1}{3} =$$ $$\frac{5}{12} - \frac{4}{12} =$$

$$\frac{8}{13} - \frac{5}{13} =$$ $$11\frac{9}{10} - 8\frac{3}{10} =$$

Section two: The student should make no more than 2 mistakes on each of these points.

Point 1: Multiply top and bottom of each these fractions by 3 to find equivalent fractions.

$$\frac{2}{5} = \text{\_\_\_\_}$$ $$\frac{1}{3} = \text{\_\_\_\_}$$ $$\frac{5}{8} = \text{\_\_\_\_}$$ $$\frac{4}{7} = \text{\_\_\_\_}$$

Point 2: Find equivalent fractions by dividing each fraction by 4.

$$\frac{4}{12} = \text{\_\_\_\_}$$ $$\frac{32}{40} = \text{\_\_\_\_}$$ $$\frac{20}{28} = \text{\_\_\_\_}$$

$$\frac{40}{48} = \text{\_\_\_\_}$$ $$\frac{12}{36} = \text{\_\_\_\_}$$ $$\frac{16}{24} = \text{\_\_\_\_}$$

Point 3: Complete this multiplication chart.

×	0	1	2	3	4	5	6	7	8	9	10	11	12
0													
1													
2													
3													
4													
5													
6													
7													
8													
9													
10													
11													
12													

Level 6 Readiness Test

This placement test assesses your student's readiness to begin *Math Lessons for a Living Education Level 6*. Please discuss any missed problems with the student in order to understand the reason that he or she missed them. Instructions for grading are at the beginning of each section. **If your student completes this test and understands the concepts, they are prepared to begin** *Math Lessons for a Living Education Level 6*.

Section One: The student should make no more than 2 mistakes on each of these points.

Point 1: Addition and Subtraction.

1.
$$\begin{array}{r} 285,230 \\ +\ 199,967 \\ \hline \end{array}$$

2.
$$\begin{array}{r} 19,002 \\ +\ 7,139 \\ \hline \end{array}$$

3.
$$\begin{array}{r} 800,045 \\ -\ 697,999 \\ \hline \end{array}$$

4.
$$\begin{array}{r} 10,000 \\ -\ 2,999 \\ \hline \end{array}$$

Point 2: Multiplication and Division.

5.
$$\begin{array}{r} 412,678 \\ \times\ 3,312 \\ \hline \end{array}$$

6.
$$\begin{array}{r} 812 \\ \times\ 88 \\ \hline \end{array}$$

7. $2\ 7\ \overline{|\ 5\ \ 6\,,\ 7\ \ 8\ \ 1}$

8. $1\ 1\ 5\ \overline{|\ 2\ \ 3\ \ 0\,,\ 5\ \ 4\ \ 2}$

Section Two: The student should make no more than 2 mistakes on each of these points.

Point 1: Story Problem. Explain and show your teacher each step.

9. A road trip is 2,540 miles long. One quarter of those miles were through mountainous terrain. Explain to your teacher how you would go about finding the number of miles that are through mountainous terrain. Write that number here:

   If you drove those miles through mountainous terrain at an average speed of 45 miles per hour, how many hours would it take you to drive through the mountainous terrain? Explain and write your answer here:

Point 2: Place Value.

Circle the digits.

10. In the tens' place: 317,002   299   512   899,982

11. In the ten thousands' place: 23,009,167   56,451   173,900

12. In the millions' place: 431,229,501   99,223,147   10,000,332

13. a. Now tell your teacher what each of the circled digits stands for.
    b. Read the numbers to your teacher.

Section Three: The student should make no more than 2 mistakes on each of these points.

Point 1: Fractions and Mixed Numbers (watch those denominators!). Explain and show.

14. $\dfrac{1}{2}$

   $+ \ \dfrac{1}{4}$

   _____

15. $6\dfrac{3}{5}$

   $- \ 2\dfrac{1}{5}$

   _____

16. $7\dfrac{6}{14}$

   $- \ 5\dfrac{2}{7}$

   _____

Point 2:

Circle the decimal or percent that matches the fraction. Explain and show your teacher as you solve each problem.

17. $\frac{1}{2}$:   40% and 0.4   20% and 0.2   50% and 0.5

18. $\frac{3}{4}$:   34% and 3.4   43% and 4.3   75% and 0.75

19. $\frac{1}{4}$:   22% and 0.22   25% and 0.25   14% and 0.14

20. $\frac{1}{5}$:   15% and 0.15   20% and 0.2   51% and 0.51

Section Four: The student should make no more than 2 mistakes on each of these points.

Point 1: Geometry.

Find the perimeter of each shape.

21.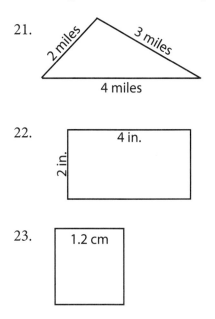

22.

    4 in.

2 in.

23.

1.2 cm

24. Find the area of the rectangle in problem 22.

25. Explain the difference between the perimeter and the area of a shape.

Ensure your student can accomplish each task on this list. **If your student can accomplish all the activities in this test, they are prepared to begin Level 1.**

☑ Know left from right

☑ Draw a straight line

☑ Trace a looping line

☑ Write their name, holding pencil correctly

☑ Follow directions successfully (i.e. play Mother May I?, giving 2-step instructions)

☑ Know colors (blue, red, yellow, orange, purple, green, brown, black, white)

☑ Use scissors correctly to cut lines at the bottom of this page

Level 2 Readiness Solutions

Instructions for grading are found at the beginning of each section. **If your student understands all the concepts on the Level 2 Placement Test, they are ready to begin** *Math Lessons for a Living Education Level 2.* Please do not place an unprepared student in this level, as it will only frustrate them and inhibit them from learning.

Part 1: The student should make no more than 2 mistakes on each of these sections.

Section one: Teacher, instruct your student to write the numbers 0–100 on the following lines. All ones' place numbers need a red underline; all tens' place numbers need a green underline; all hundreds' place numbers need a blue underline.

0	1	2	3	4	5	6	7	8	9	10
11	12	13	14	15	16	17	18	19	20	
21	22	23	24	25	26	27	28	29	30	
31	32	33	34	35	36	37	38	39	40	
41	42	43	44	45	46	47	48	49	50	
51	52	53	54	55	56	57	58	59	60	
61	62	63	64	65	66	67	68	69	70	
71	72	73	74	75	76	77	78	79	80	
81	82	83	84	85	86	87	88	89	90	
91	92	93	94	95	96	97	98	99	100	

Part 2: Teacher, instruct your student to underline every number on the previous page that is in the ones' place with a red crayon/pencil, every number in the tens' place with a green crayon/pencil, every number in the hundreds' place with a blue crayon/pencil.

Orally, have your student answer these questions.

☐   In the number 236, what does 6 stand for? (b)
a) six groups of ten
b) six groups of one
c) six groups of one hundred

☐   In the number 236, what does 3 stand for?   (a)
a) three groups of ten
b) three groups of one
c) three groups of one hundred

☐   In the number 236, what does 2 stand for? (c)
a) two groups of ten
b) two groups of one
c) two groups of one hundred

Section two: The student should make no more than 1 mistake on each of these points.

Point 1: Teacher have your student draw hands on these clocks to show the correct time.

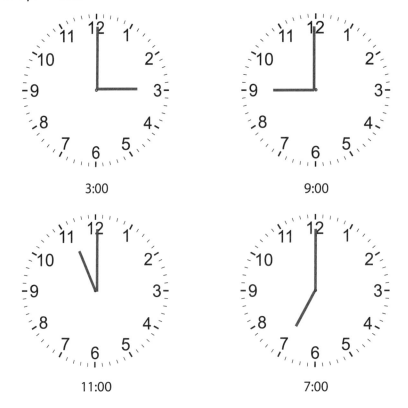

3:00

9:00

11:00

7:00

Point 2: Teacher, have your student answer these. They should do these from memory; watch them carefully and take note of the ones they have to think or count to answer. (This is about seeing if your student understands the concept of addition. If they can answer from memory, this is a plus, but not absolutely necessary.)

$$2 + 8 = 10 \qquad 4 + 2 = 6 \qquad 4 + 5 = 9$$

$$3 + 2 = 5 \qquad 9 + 1 = 10 \qquad 4 + 4 = 8$$

$$6 + 3 = 9 \qquad 2 + 3 = 5 \qquad 8 + 2 = 10$$

Point 3: Have your student answer these quickly. They should do these from memory; watch them carefully and take note of the ones they have to think or count to answer. (This is about seeing if your student understands the concept of subtraction. If they can answer from memory, this is a plus, but not absolutely necessary.)

$$10 - 2 = 8 \qquad 8 - 3 = 5 \qquad 6 - 2 = 4$$

$$9 - 7 = 2 \qquad 10 - 5 = 5 \qquad 9 - 5 = 4$$

$$10 - 8 = 2 \qquad 7 - 4 = 3 \qquad 6 - 5 = 1$$

Section three: The student should make no more than 1 mistake on each of these.
Teacher, have your student narrate to you the relationship between addition and subtraction. Do not help or coach your student at all. It is extremely important that they understand the relationship between these two operations. If your student has done well on the other parts of this placement test, but does not understand this concept of relationship, please take a few minutes to use manipulatives to show them with the hands-on/visual/auditory approach. If they are not understanding this concept easily and are not able to narrate back to you as they show you with the manipulatives, they are not ready for *Math Level 2.*

Level 3 Readiness Solutions

Instructions for grading are found at the beginning of each section. **If your student understands all the concepts on the Level 3 Placement Test, they are ready to begin** *Math Lessons for a Living Education Level 3*. Please do not place an unprepared student in this level, as it will only frustrate them and inhibit them from learning.

Section one: The student should make no more than 2 mistakes on each of these points.

Point 1: Fill in the chart correctly.

	Thousands	Hundreds	Tens	Ones
6,011	6	0	1	1
792		7	9	2
4,009	4	0	0	9
8,178	8	1	7	8
2,060	2	0	6	0

Point 2: Look at the numbers in the chart above. Color each even number green. Color each odd number blue.

Point 3: What numbers do odd numbers end in? <u>1, 3, 5, 7, 9</u>

What numbers do even numbers end in? <u>2, 4, 6, 8, 0</u>

Section two: The student should make no more than 2 mistakes on each of these points.

Point 1: Write the correct time shown on each clock.

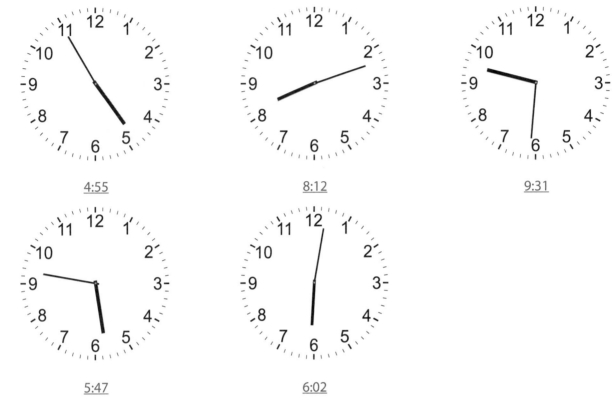

4:55

8:12

9:31

5:47

6:02

Point 2: Count the money and write the correct amount.

$4.20

$10.03

$12.65

$101.99

Point 3: Find the perimeter of each shape.

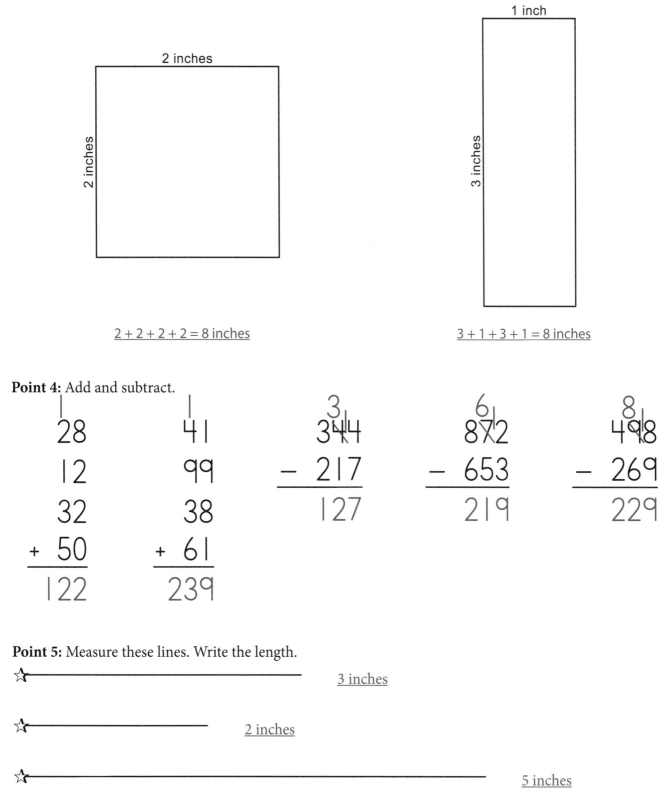

$2 + 2 + 2 + 2 = 8$ inches $3 + 1 + 3 + 1 = 8$ inches

Point 4: Add and subtract.

$$
\begin{array}{r}
28 \\
12 \\
32 \\
+\ 50 \\
\hline
122
\end{array}
\qquad
\begin{array}{r}
41 \\
99 \\
38 \\
+\ 61 \\
\hline
239
\end{array}
\qquad
\begin{array}{r}
3\,4\,4 \\
-\ 2\,1\,7 \\
\hline
1\,2\,7
\end{array}
\qquad
\begin{array}{r}
8\,7\,2 \\
-\ 6\,5\,3 \\
\hline
2\,1\,9
\end{array}
\qquad
\begin{array}{r}
4\,9\,8 \\
-\ 2\,6\,9 \\
\hline
2\,2\,9
\end{array}
$$

Point 5: Measure these lines. Write the length.

☆ ———————————————   <u>3 inches</u>

☆ ——————————   <u>2 inches</u>

☆ ———————————————————————   <u>5 inches</u>

Level 4 Readiness Solutions

Instructions for grading are found at the beginning of each section. **If your student understands all the concepts on the Level 4 Placement Test, they are ready to begin** *Math Lessons for a Living Education Level 4*. Please do not place an unprepared student in this level, as it will only frustrate them and inhibit them from learning.

Section one: The student should make no more than 2 mistakes on each of these points.

Point 1: Add and subtract.

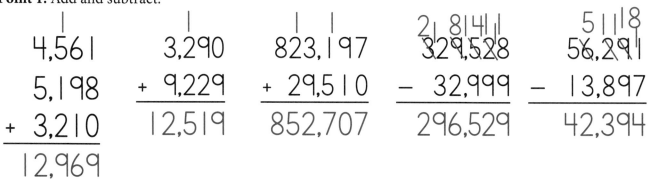

Point 2: Round these numbers to the nearest 10.

23	20
891	890
466	470
138	140

Round these numbers to the nearest 100.

189	200
2,345	2,300
982	1,000
312	300

Round these numbers to the nearest 1,000.

3,780	4,000
12,428	12,000
9,621	10,000
13,289	13,000

Point 3: Complete this multiplication chart.

×	1	2	3	4	5	6	7	8	9	10
1	1	2	3	4	5	6	7	8	9	10
2	2	4	6	8	10	12	14	16	18	20
3	3	6	9	12	15	18	21	24	27	30
4	4	8	12	16	20	24	28	32	36	40
5	5	10	15	20	25	30	35	40	45	50
6	6	12	18	24	30	36	42	48	54	60
7	7	14	21	28	35	42	49	56	63	70
8	8	16	24	32	40	48	56	64	72	80
9	9	18	27	36	45	54	63	72	81	90
10	10	20	30	40	50	60	70	80	90	100

Point 4: Narrate to your teacher the relationship between multiplication and division. Use manipulatives to demonstrate your understanding.

(Note to the teacher: this point is a make or break. If your student does not understand multiplication and division well enough to confidently and clearly narrate to you the relationship between multiplication and division, seriously consider placing them in the previous level in this series.)

Section two: The student should make no more than 2 mistakes on each of these points.

Point 1: Find the area. Write the equations for each one.

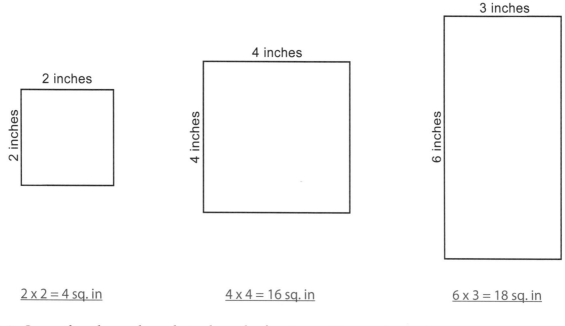

2 inches

2 inches

4 inches

4 inches

3 inches

6 inches

2 x 2 = 4 sq. in   4 x 4 = 16 sq. in   6 x 3 = 18 sq. in

Point 2: Correctly color each circle to show the fraction written under each one.

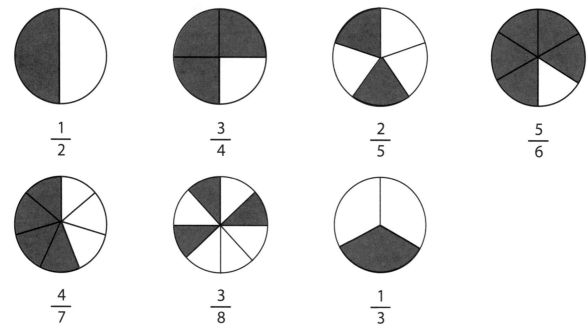

$\frac{1}{2}$ $\frac{3}{4}$ $\frac{2}{5}$ $\frac{5}{6}$

$\frac{4}{7}$ $\frac{3}{8}$ $\frac{1}{3}$

Point 3: Solve these word problems.

There were 32 tulips at the flower stand. If 4 ladies bought an equal number of the tulips, how many tulips did they each buy?

$$32 \div 4 = 8$$

The family drove 126 miles before lunch. After lunch, they drove 253. How many more miles did they drive in the afternoon than in the morning? How many miles did they drive in the morning and the afternoon together?

$$253 - 126 = 127 \;/\; 126 + 253 = 379$$

Point 4: Solve these problems.

Circle groups of 3. $\frac{1}{6}$ of 18 = <u>3</u>   6 x <u>3</u> = 18

Circle groups of 5. $\frac{1}{2}$ of 10 = <u>5</u>   2 x <u>5</u> = 10

Circle groups of 4. $\frac{1}{3}$ of 12 = <u>4</u>   3 x <u>4</u> = 12

Point 5: Write the Roman Numeral for each number.

1	I
5	V
10	X
50	L
100	C
1,000	M

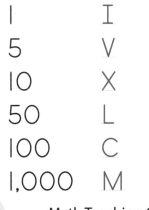

114   Math Teaching Companion

Level 5 Readiness Solutions

Instructions for grading are found at the beginning of each section. **If your student understands all the concepts on the Level 5 Placement Test, they are ready to begin** *Math Lessons for a Living Education Level 5*. Please do not place an unprepared student in this level, as it will only frustrate them and inhibit them from learning.

Section one: The student should make no more than 2 mistakes on each of these points.

Point 1: Add and subtract.

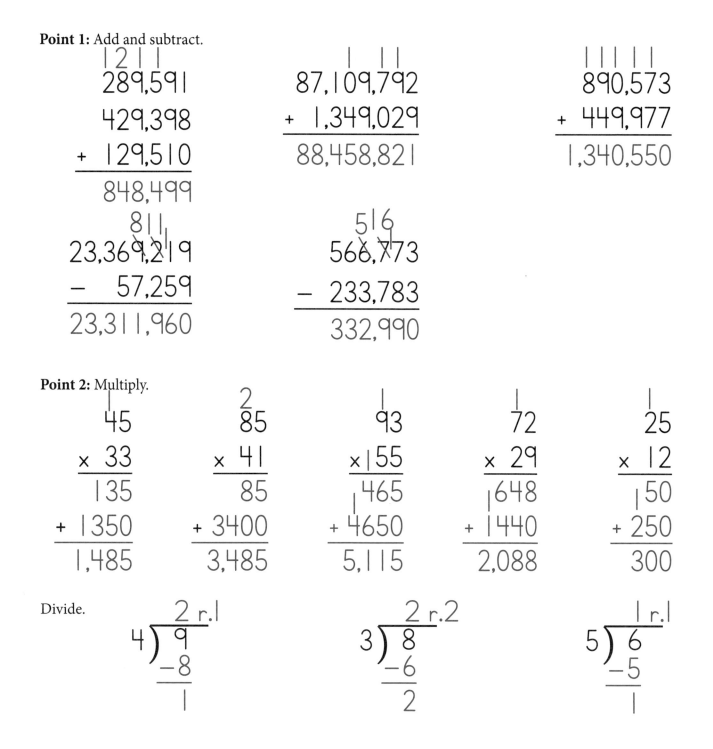

$$
\begin{array}{r}
\overset{1}{2}\overset{2}{1}\overset{1}{1}\ \\
289{,}591 \\
429{,}398 \\
+\ 129{,}510 \\
\hline
848{,}499
\end{array}
\qquad
\begin{array}{r}
\overset{1}{\ }\ \overset{1}{\ }\overset{1}{\ } \\
87{,}109{,}792 \\
+\ 1{,}349{,}029 \\
\hline
88{,}458{,}821
\end{array}
\qquad
\begin{array}{r}
\overset{1}{1}\overset{1}{1}\overset{1}{1}\ \overset{1}{1}\overset{1}{1} \\
890{,}573 \\
+\ 449{,}977 \\
\hline
1{,}340{,}550
\end{array}
$$

$$
\begin{array}{r}
\overset{8}{2}\overset{1}{3}{,}36\overset{1}{9}{,}\overset{1}{2}19 \\
-\ \ \ 57{,}259 \\
\hline
23{,}311{,}960
\end{array}
\qquad
\begin{array}{r}
\overset{5}{5}\overset{1}{6}\overset{6}{8}{,}\cancel{7}73 \\
-\ 233{,}783 \\
\hline
332{,}990
\end{array}
$$

Point 2: Multiply.

$$
\begin{array}{r}
\overset{1}{45} \\
\times\ 33 \\
\hline
135 \\
+\ 1350 \\
\hline
1{,}485
\end{array}
\quad
\begin{array}{r}
\overset{2}{85} \\
\times\ 41 \\
\hline
85 \\
+\ 3400 \\
\hline
3{,}485
\end{array}
\quad
\begin{array}{r}
\overset{1}{93} \\
\times\ 55 \\
\hline
465 \\
+\ 4650 \\
\hline
5{,}115
\end{array}
\quad
\begin{array}{r}
\overset{1}{72} \\
\times\ 29 \\
\hline
648 \\
+\ 1440 \\
\hline
2{,}088
\end{array}
\quad
\begin{array}{r}
\overset{1}{25} \\
\times\ 12 \\
\hline
50 \\
+\ 250 \\
\hline
300
\end{array}
$$

Divide.

$$
\begin{array}{r}
2\ \text{r.}1 \\
4\overline{)\,9} \\
-8 \\
\hline
1
\end{array}
\qquad
\begin{array}{r}
2\ \text{r.}2 \\
3\overline{)\,8} \\
-6 \\
\hline
2
\end{array}
\qquad
\begin{array}{r}
1\ \text{r.}1 \\
5\overline{)\,6} \\
-5 \\
\hline
1
\end{array}
$$

Point 3: Word Problems. The toy shop had 2,872 boomerangs in stock for the Christmas sale. After the sale, there were 1,988 boomerangs still in stock. The store decided to place half of the boomerangs on the clearance shelves, and donate the other half to a missions organization. How many boomerangs were donated to the missions organization?

$$1,988 ÷ 2 = 994 \text{ donated to missions}$$

When the boomerangs were delivered to the missions organization, they were equally packaged in two large boxes. How many were in each box? When the workers at the organization opened one of the boxes, they found that a dozen boomerangs had been damaged in the shipment. How many boomerangs were undamaged in that box?

$$994 ÷ 2 = 497 \text{ in each box} / 497 − 12 = 485 \text{ undamaged}$$

Point 4: Add and subtract these fractions.

$$\frac{3}{7} + \frac{2}{7} = \frac{5}{7} \qquad 2\frac{2}{5} + 1\frac{1}{5} = 3\frac{3}{5} \qquad 3\frac{5}{9} + 2\frac{1}{9} = 5\frac{6}{9}$$

$$\frac{3}{11} + \frac{6}{11} = \frac{9}{11} \qquad 6\frac{2}{3} − 4\frac{1}{3} = 2\frac{1}{3} \qquad \frac{5}{12} − \frac{4}{12} = \frac{1}{12}$$

$$\frac{8}{13} − \frac{5}{13} = \frac{3}{13} \qquad 11\frac{9}{10} − 8\frac{3}{10} = 3\frac{6}{10}$$

Section two: The student should make no more than 2 mistakes on each of these points.

Point 1: Multiply top and bottom of each these fractions by 3 to find equivalent fractions.

$$\frac{2}{5} = \frac{6}{15} \qquad \frac{1}{3} = \frac{3}{9} \qquad \frac{5}{8} = \frac{15}{24} \qquad \frac{4}{7} = \frac{12}{21}$$

Point 2: Find equivalent fractions by dividing each fraction by 4.

$$\frac{4}{12} = \frac{1}{3} \qquad \frac{32}{40} = \frac{8}{10} \qquad \frac{20}{28} = \frac{5}{7}$$

$$\frac{40}{48} = \frac{10}{12} \qquad \frac{12}{36} = \frac{3}{9} \qquad \frac{16}{24} = \frac{4}{6}$$

Point 3: Complete this multiplication chart.

×	0	1	2	3	4	5	6	7	8	9	10	11	12
0	0	0	0	0	0	0	0	0	0	0	0	0	0
1	0	1	2	3	4	5	6	7	8	9	10	11	12
2	0	2	4	6	8	10	12	14	16	18	20	22	24
3	0	3	6	9	12	15	18	21	24	27	30	33	36
4	0	4	8	12	16	20	24	28	32	36	40	44	48
5	0	5	10	15	20	25	30	35	40	45	50	55	60
6	0	6	12	18	24	30	36	42	48	54	60	66	72
7	0	7	14	21	28	35	42	49	56	63	70	77	84
8	0	8	16	24	32	40	48	56	64	72	80	88	96
9	0	9	18	27	36	45	54	63	72	81	90	99	108
10	0	10	20	30	40	50	60	70	80	90	100	110	120
11	0	11	22	33	44	55	66	77	88	99	110	121	132
12	0	12	24	36	48	60	72	84	96	108	120	132	144

Level 6 Readiness Solutions

Instructions for grading are found at the beginning of each section. **If your student understands all the concepts on the Level 6 Placement Test, they are ready to begin** *Math Lessons for a Living Education Level 6*. Please do not place an unprepared student in this level, as it will only frustrate them and inhibit them from learning.

Section One: The student should make no more than 2 mistakes on each of these points.

Point 1: Addition and Subtraction.

1.
$$\begin{array}{r} {}^{111} \\ 285{,}230 \\ +\ \ 199{,}967 \\ \hline 485{,}197 \end{array}$$

2.
$$\begin{array}{r} {}^{1\ \ \ 1} \\ 19{,}002 \\ +\ \ \ 7{,}139 \\ \hline 26{,}141 \end{array}$$

3.
$$\begin{array}{r} {}^{799\ 913} \\ 800{,}045 \\ -\ \ 697{,}999 \\ \hline 102{,}046 \end{array}$$

4.
$$\begin{array}{r} {}^{9\ \ 99} \\ 10{,}000 \\ -\ \ \ 2{,}999 \\ \hline 7{,}001 \end{array}$$

Point 2: Multiplication and Division.

5.
$$\begin{array}{r} 412{,}678 \\ \times\ \ \ \ \ \ 3{,}312 \\ \hline 825356 \\ 4126780 \\ 123803400 \\ +1238034000 \\ \hline 1{,}366{,}789{,}536 \end{array}$$

6.
$$\begin{array}{r} 812 \\ \times\ \ \ \ 88 \\ \hline 6496 \\ +\ 64960 \\ \hline 71{,}456 \end{array}$$

7.
```
            2 , 1   0   3
  2 7 | 5   6 , 7   8   1
      - 5   4
        2   7
      - 2   7
            0   8   1
              - 8   1
                    0
```

8.
```
                  2 , 0   0   4  R.82
  1 1 5 | 2   3   0 , 5   4   2
        - 2   3   0
                0   ⁸ ¹4   2
                  - 4   6   0
                        8   2
```

Section Two: The student should make no more than 2 mistakes on each of these points.

Point 1: Story Problem. Explain and show your teacher every step of this story problem.

9. A road trip is 2,540 miles long. One quarter of those miles were through mountainous terrain. Explain to your teacher how you would go about finding the number of miles that are through mountainous terrain. Write that number here: 635

$$
\begin{array}{r}
6\ \ 3\ \ 5 \\
4\ \overline{)\ 2\ ,5\ \ 4\ \ 0}
\end{array}
$$

If you drove those miles through mountainous terrain at an average speed of 45 miles per hour, how many hours would it take you to drive through the mountainous terrain? Explain and write your answer here: 14

$$
\begin{array}{r}
1\ \ 4 \\
4\ \ 5\ \overline{)\ 6\ \ 3\ \ 5} \\
-\ 4\ \ 5\ \ \ \ \\
\overline{\ \ \ 1\ \ 8\ \ 5} \\
-\ 1\ \ 8\ \ 0 \\
\overline{\ \ \ \ \ \ \ \ 5}
\end{array}
$$

Point 2: Place Value

Circle the digits.

10. In the tens' place: 317,00②   2⑨9   5①2   899,9⑧2

11. In the ten-thousands' place: 23,0⓪9,167   ⑤6,451   1⑦3,900

12. In the millions' place: 43①,229,501   9⑨,223,147   1⓪,000,332

13. a. Now tell your teacher what each of the circled digits stands for.
    b. Read the numbers to your teacher.

Section Three: The student should make no more than 2 mistakes on each of these points.

Point 1: Fractions and Mixed Numbers (watch those denominators!). Explain and show.

14.
$$
\begin{array}{r}
\frac{1}{2} \quad \frac{2}{4} \\
+\ \frac{1}{4} \ +\ \frac{1}{4} \\
\hline
\frac{3}{4}
\end{array}
$$

15.
$$
\begin{array}{r}
6\frac{3}{5} \\
-\ 2\frac{1}{5} \\
\hline
4\frac{2}{5}
\end{array}
$$

16.
$$
\begin{array}{r}
7\frac{6}{14} \quad 7\frac{6}{14} \\
-\ 5\frac{2}{7} \ -\ 5\frac{4}{14} \\
\hline
2\frac{2}{14} \quad 2\frac{2}{14}
\end{array}
$$

Point 2:
Circle the decimal or percent that matches the fraction. Explain and show your teacher as you solve each problem.

17. $\frac{1}{2}$: 40% and 0.4 20% and 0.2 ⟨50% and 0.5⟩

18. $\frac{3}{4}$: 34% and 3.4 43% and 4.3 ⟨75% and 0.75⟩

19. $\frac{1}{4}$: 22% and 0.22 ⟨25% and 0.25⟩ 14% and 0.14

20. $\frac{1}{5}$: 15% and 0.15 ⟨20% and 0.2⟩ 51% and 0.51

Section Four: The student should make no more than 2 mistakes on each of these points.

Point 1: Geometry.

Find the perimeter of each shape.

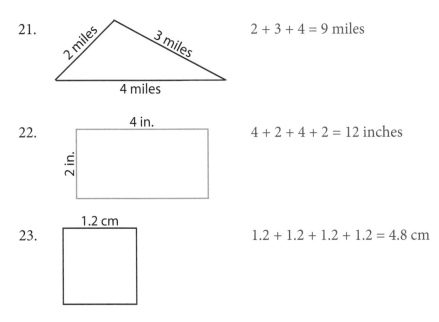

21. 2 + 3 + 4 = 9 miles

22. 4 + 2 + 4 + 2 = 12 inches

23. 1.2 + 1.2 + 1.2 + 1.2 = 4.8 cm

24. Find the area of the rectangle in problem 22.
    2 x 4 = 8 square inches

25. Explain the difference between the perimeter and the area of a shape.
    Area: the measurement of the inside of a shape.
    Perimeter: the distance around (or outside of) a shape.

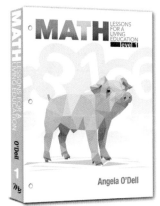
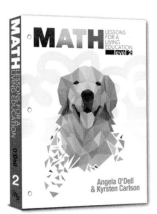
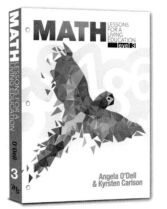

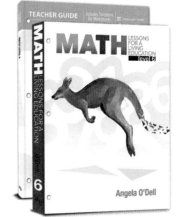